A Short Course in Differential Topology

Manifolds abound in mathematics and physics, and increasingly in cybernetics and visualization, where they often reflect properties of complex systems and their configurations. Differential topology gives us tools to study these spaces and extract information about the underlying systems.

This book offers a concise and modern introduction to the core topics of differential topology for advanced undergraduates and beginning graduate students. It covers the basics on smooth manifolds and their tangent spaces before moving on to regular values and transversality, smooth flows and differential equations on manifolds, and the theory of vector bundles and locally trivial fibrations. The final chapter gives examples of local-to-global properties, a short introduction to Morse theory and a proof of Ehresmann's fibration theorem.

The treatment is hands-on, including many concrete examples and exercises woven into the text, with hints provided to guide the student.

Bjørn Ian Dundas is Professor in the Mathematics Department at the University of Bergen. Besides his research and teaching he is the author of three books. When not doing mathematics he enjoys hiking and fishing from his rowboat in northern Norway.

CAMBRIDGE MATHEMATICAL TEXTBOOKS

Cambridge Mathematical Textbooks is a program of undergraduate and beginning graduate-level textbooks for core courses, new courses, and interdisciplinary courses in pure and applied mathematics. These texts provide motivation with plenty of exercises of varying difficulty, interesting examples, modern applications, and unique approaches to the material.

Advisory Board

John B. Conway, *George Washington University*
Gregory F. Lawler, *University of Chicago*
John M. Lee, *University of Washington*
John Meier, *Lafayette College*
Lawrence C. Washington, *University of Maryland, College Park*

A complete list of books in the series can be found at www.cambridge.org/mathematics. Recent titles include the following:

A Short Course in Differential Topology

Bjørn Ian Dundas

Universitetet i Bergen, Norway

CAMBRIDGE
UNIVERSITY PRESS

CAMBRIDGE
UNIVERSITY PRESS

University Printing House, Cambridge CB2 8BS, United Kingdom

One Liberty Plaza, 20th Floor, New York, NY 10006, USA

477 Williamstown Road, Port Melbourne, VIC 3207, Australia

314-321, 3rd Floor, Plot 3, Splendor Forum, Jasola District Centre, New Delhi - 110025, India

103 Penang Road, #05-06/07, Visioncrest Commercial, Singapore 238467

Cambridge University Press is part of the University of Cambridge.

It furthers the University's mission by disseminating knowledge in the pursuit of education, learning and research at the highest international levels of excellence.

www.cambridge.org
Information on this title: www.cambridge.org/9781108425797
DOI: 10.1017/9781108349130

First published 2018

A catalogue record for this publication is available from the British Library

Library of Congress Cataloging in Publication data
Names: Dundas, B. I. (Bjorn Ian), author.
Title: A short course in differential topology / Bjorn Ian Dundas
    (Universitetet i Bergen, Norway).
Other titles: Cambridge mathematical textbooks.
Description: Cambridge, United Kingdom ; New York, NY :
    Cambridge University Press, 2018. | Series: Cambridge mathematical textbooks
Identifiers: LCCN 2017061465| ISBN 9781108425797 (hardback ; alk. paper) |
    ISBN 1108425798 (hardback ; alk. paper)
Subjects: LCSH: Differential topology.
Classification: LCC QA613.6 .D86 2018 | DDC 514/.72–dc23
LC record available at https://lccn.loc.gov/2017061465

ISBN 978-1-108-42579-7 Hardback

Contents

Preface

In his inaugural lecture in 1854[1], Riemann introduced the concept of an "n-fach ausgedehnte Grösse" – roughly something that has "n degrees of freedom" and which we now would call an n-dimensional manifold.

Examples of manifolds are all around us and arise in many applications, but formulating the ideas in a satisfying way proved to be a challenge inspiring the creation of beautiful mathematics. As a matter of fact, much of the mathematical language of the twentieth century was created with manifolds in mind.

Modern texts often leave readers with the feeling that they are getting the answer before they know there is a problem. Taking the historical approach to this didactic problem has several disadvantages. The pioneers were brilliant mathematicians, but still they struggled for decades getting the concepts right. We must accept that we are standing on the shoulders of giants.

The only remedy I see is to give carefully chosen examples to guide the mind to ponder over the questions that you would actually end up wondering about even after spending a disproportionate amount of time. In this way I hope to encourage readers to appreciate and internalize the solutions when they are offered.

These examples should be concrete. On the other end of the scale, proofs should also be considered as examples: they are examples of successful reasoning. "Here is a way of handling such situations!" However, no amount of reading can replace doing, so there should be many opportunities for trying your hand.

In this book I have done something almost unheard of: I provide (sometimes quite lengthy) hints for all the exercises. This requires quite a lot of self-discipline from the reader: it is *very* hard not to peek at the solution too early. There are several reasons for including hints. First and foremost, the exercises are meant to be an integral part of class life. The exercises can be assigned to students who present their solutions in problem sessions, in which case the students must internalize their solution, but at the same time should be offered some moral support to lessen the social stress. Secondly, the book was designed for students who – even if eager to learn – are in need of more support with respect to how one can reason about the material. Trying your hand on the problem, getting stuck, taking a peek to see whether you glimpse an idea, trying again ... and eventually getting a solution that you believe in and which you can discuss in class is *way* preferable to not having

---

[1] https://en.wikipedia.org/wiki/Bernhard_Riemann

anything to bring to class. A side effect is that this way makes it permissible to let the students develop parts of the text themselves without losing accountability. Lastly, though this was not a motivation for me, providing hints makes the text better suited for self-study.

Why This Book?

The year I followed the manifold course (as a student), we used Spivak [20], and I came to love the "Great American Differential Geometry book". At the same time, I discovered a little gem by Bröker and Jänich [4] in the library that saved me on some of the occasions when I got totally befuddled. I spent an inordinate amount of time on that class.

Truth be told, there are many excellent books on manifolds out there; to name just three, Lee's book [13] is beautiful; in a macho way so is Kosinski's [11]; and Milnor's pearl [15] will take you all the way from zero to framed cobordisms in 50 pages. Why write one more?

Cambridge University Press wanted "A Short Introduction" to precede my original title "Differential Topology". They were right: this is a far less ambitious text than the ones I have mentioned, and was designed for the students who took my classes. As a student I probably could provide a proof for all the theorems, but if someone asked me to check a very basic fact like "Is this map smooth?" I would feel that it was so for "obvious reasons" and hope for the life of me that no one would ask "why?" The book offers a modern framework while not reducing everything to some sort of magic. This allows us to take a hands-on approach; we are less inclined to identify objects without being specific about *how* they should be identified, removing some of the anxiety about "variables" and "coordinates changing" this or that way.

Spending time on the basics but still aiming at a one-semester course forces some compromises on this fairly short book. Sadly, topics like Sard's theorem, Stokes' theorem, differential forms, de Rham cohomology, differential equations, Riemannian geometry and surfaces, imbedding theory, K-theory, singularities, foliations and analysis on manifolds are barely touched upon.

At the end of the term, I hope that the reader will have internalized the fundamental ideas and will be able to use the basic language and tools with enough confidence to apply them in other fields, and to embark on more ambitious texts. Also, I wanted to prove Ehresmann's fibration theorem because I think it is cool.

How to Start Reading

The core curriculum consists of Chapters 2–8. The introduction in Chapter 1 is not strictly necessary for highly motivated readers who cannot wait to get to the theory, but provides some informal examples and discussions meant to put the later material into some perspective. If you are weak on point set topology, you will probably want to read Appendix A in parallel with Chapter 2. You should also be aware of the fact that Chapters 4 and 5 are largely independent, and, apart from a few exercises, can be read in any order. Also, at the cost of removing some exercises and examples, the sections on derivations (Section 3.5), orientations (Section 6.7), the generalized Gauss map (Section 6.8), second-order differential equations (Section 7.4), the exponential map (Section 8.2.7) and Morse theory (Section 8.4) can be removed from the curriculum without disrupting the logical development of ideas. The cotangent space/bundle material (Sections 3.4 and 5.6) can be omitted at the cost of using the dual tangent bundle from Chapter 6 onward.

Do the exercises, and only peek(!) at the hints if you *really* need to.

Prerequisites

Apart from relying on standard courses in multivariable analysis and linear algebra, this book is designed for readers who have already completed either a course in analysis that covers the basics of metric spaces or a first course in general topology. Most students will feel that their background in linear algebra could have been stronger, but it is to be hoped that seeing it used will increase their appreciation of things beyond Gaussian elimination.

Acknowledgments

First and foremost, I am indebted to the students and professors who have used the online notes and given me invaluable feedback. Special thanks go to Håvard Berland, Elise Klaveness, Torleif Veen, Karen Sofie Rønæss, Ivan Viola, Samuel Littig, Eirik Berge, Morten Brun and Andreas Leopold Knutsen. I owe much to a couple of anonymous referees (some unsolicited, but all very helpful) for their diligent reading and many constructive comments. They were probably right to insist that the final version should not actively insult the reader even if it means adopting less colorful language. The people at Cambridge University Press have all been very helpful, and I want especially to thank Clare Dennison, Tom Harris and Steven Holt.

My debt to the books [8], [11], [12], [13], [15], [14], [20] and in particular [4] should be evident from the text.

I am grateful to UiB for allowing me to do several revisions in an inspiring environment (Kistrand, northern Norway), and to the Hausdorff Institute in Bonn and the University of Copenhagen for their hospitality. The frontispiece is an adaption of one of my T-shirts. Thanks to Vår Iren Hjorth Dundas.

Notation

We let $\mathbf{N} = \{0, 1, 2, \ldots\}$, $\mathbf{Z} = \{\ldots, -1, 0, 1, \ldots\}$, $\mathbf{Q}$, $\mathbf{R}$ and $\mathbf{C}$ be the sets of natural numbers, integers, rational numbers, real numbers and complex numbers. If X and Y are two sets, $X \times Y$ is the set of ordered pairs (x, y) with x an element in X and y an element in Y. If n is a natural number, we let $\mathbf{R}^n$ and $\mathbf{C}^n$ be the vector spaces of ordered n-tuples of real and complex numbers. Occasionally we may identify $\mathbf{C}^n$ with $\mathbf{R}^{2n}$. If $p = (p_1, \ldots, p_n) \in \mathbf{R}^n$, we let $|p|$ be the norm $\sqrt{p_1^2 + \cdots + p_n^2}$. The sphere of dimension n is the subset $S^n \subseteq \mathbf{R}^{n+1}$ of all $p = (p_0, \ldots, p_n) \in \mathbf{R}^{n+1}$ with $|p| = 1$ (so that $S^0 = \{-1, 1\} \subseteq \mathbf{R}$, and S^1 can be viewed as all the complex numbers $e^{i\theta}$ of unit length).

Given functions $f \colon X \to Y$ and $g \colon Y \to Z$, we write gf for the composite, and $g \circ f$ only if the notation is cluttered and the $\circ$ improves readability. The constellation $g \cdot f$ will occur in the situation where f and g are functions with the same source and target, and where multiplication makes sense in the target. If X and Y are topological spaces, a continuous function $f \colon X \to Y$ is simply referred to as a *map*.

1 Introduction

The earth is round. This may at one point have been hard to believe, but we have grown accustomed to it even though our everyday experience is that the earth is (fairly) flat. Still, the most effective way to illustrate it is by means of maps: a globe (Figure 1.1) is a very neat device, but its global(!) character makes it less than practical if you want to represent fine details.

This phenomenon is quite common: locally you can represent things by means of "charts", but the global character can't be represented by a single chart. You need an entire atlas, and you need to know how the charts are to be assembled, or, even better, the charts overlap so that we know how they all fit together. The mathematical framework for working with such situations is manifold theory. Before we start off with the details, let us take an informal look at some examples illustrating the basic structure.

1.1 A Robot's Arm

To illustrate a few points which will be important later on, we discuss a concrete situation in some detail. The features that appear are special cases of general phenomena, and the example should provide the reader with some déjà vu experiences later on, when things are somewhat more obscure.

Figure 1.1. A globe. Photo by DeAgostini/Getty Images.

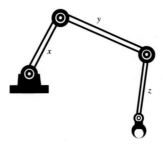

Figure 1.2.

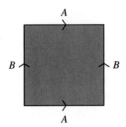

Figure 1.3.

Consider a robot's arm. For simplicity, assume that it moves in the plane, and has three joints, with a telescopic middle arm (see Figure 1.2).

Call the vector defining the inner arm x, that for the second arm y and that for the third arm z. Assume $|x| = |z| = 1$ and $|y| \in [1, 5]$. Then the robot can reach anywhere inside a circle of radius 7. But most of these positions can be reached in several different ways.

In order to control the robot optimally, we need to understand the various configurations, and how they relate to each other.

As an example, place the robot at the origin and consider all the possible positions of the arm that reach the point $P = (3, 0) \in \mathbf{R}^2$, i.e., look at the set T of all triples $(x, y, z) \in \mathbf{R}^2 \times \mathbf{R}^2 \times \mathbf{R}^2$ such that

$$x + y + z = (3, 0), \qquad |x| = |z| = 1, \qquad |y| \in [1, 5].$$

We see that, under the restriction $|x| = |z| = 1$, x and z can be chosen arbitrarily, and determine y uniquely. So T is "the same as" the set

$$\{(x, z) \in \mathbf{R}^2 \times \mathbf{R}^2 \mid |x| = |z| = 1\}.$$

Seemingly, our space T of configurations resides in four-dimensional space $\mathbf{R}^2 \times \mathbf{R}^2 \cong \mathbf{R}^4$, but that is an illusion – the space is two-dimensional and turns out to be a familiar shape. We can parametrize x and z by angles if we remember to identify the angles 0 and 2π. So T is what you get if you consider the square $[0, 2\pi] \times [0, 2\pi]$ and identify the edges as in Figure 1.3. See

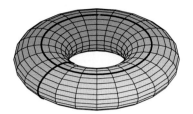

Figure 1.4.

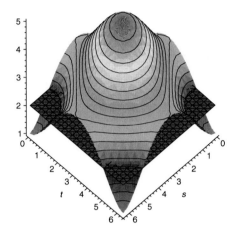

Figure 1.5.

www.it.brighton.ac.uk/staff/jt40/MapleAnimations/Torus.html for a nice animation of how the plane model gets glued.

In other words, the set T of all positions such that the robot reaches $P = (3, 0)$ may be identified with the torus in Figure 1.4. This is also true topologically in the sense that "close configurations" of the robot's arm correspond to points close to each other on the torus.

1.1.1 Question

What would the space S of positions look like if the telescope got stuck at $|y| = 2$?

Partial answer to the question: since $y = (3, 0) - x - z$ we could try to get an idea of what points of T satisfy $|y| = 2$ by means of inspection of the graph of $|y|$. Figure 1.5 is an illustration showing $|y|$ as a function of T given as a graph over $[0, 2\pi] \times [0, 2\pi]$, and also the plane $|y| = 2$.

The desired set S should then be the intersection shown in Figure 1.6. It looks a bit weird before we remember that the edges of $[0, 2\pi] \times [0, 2\pi]$ should be identified. On the torus it looks perfectly fine; and we can see this if we change our perspective a bit. In order to view T we chose $[0, 2\pi] \times [0, 2\pi]$ with identifications along the boundary. We could just as well have chosen $[-\pi, \pi] \times [-\pi, \pi]$, and then the picture would have looked like Figure 1.7. It does not touch the boundary,

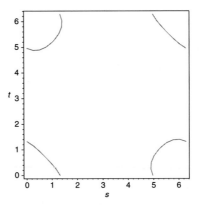

Figure 1.6.

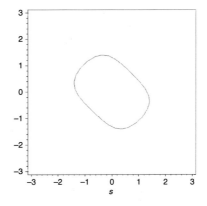

Figure 1.7.

so we do not need to worry about the identifications. As a matter of fact, S is homeomorphic to the circle (*homeomorphic* means that there is a bijection between S and the circle, and both the function from the circle to S and its inverse are continuous. See Definition A.2.8).

1.1.2 Dependence on the Telescope's Length

Even more is true: we notice that S looks like a smooth and nice curve. This will not happen for all values of $|y|$. The exceptions are $|y| = 1$, $|y| = 3$ and $|y| = 5$. The values 1 and 5 correspond to one-point solutions. When $|y| = 3$ we get a picture like Figure 1.8 (the solution really ought to touch the boundary).

We will learn to distinguish between such circumstances. They are qualitatively different in many aspects, one of which becomes apparent if we view the example shown in Figure 1.9 with $|y| = 3$ with one of the angles varying in $[0, 2\pi]$ while the other varies in $[-\pi, \pi]$. With this "cross" there is no way our solution space is homeomorphic to the circle. You can give an interpretation of the picture

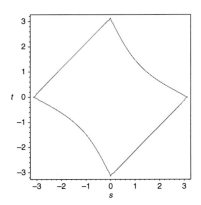

Figure 1.8.

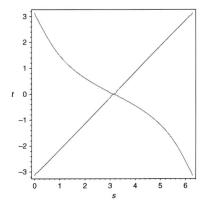

Figure 1.9.

above: the straight line is the movement you get if you let $x = z$ (like two wheels of equal radius connected by a coupling rod y on an old-fashioned train), whereas the curved line corresponds to x and z rotating in opposite directions (very unhealthy for wheels on a train).

Actually, this cross comes from a "saddle point" in the graph of $|y|$ as a function of T: it is a "critical" value at which all sorts of bad things can happen.

1.1.3 Moral

The configuration space T is smooth and nice, and we get different views on it by changing our "coordinates". By considering a function on T (in our case the length of y) and restricting to the subset of T corresponding to a given value of our function, we get qualitatively different situations according to what values we are looking at. However, away from the "critical values" we get smooth and nice subspaces, see in particular Theorem 4.4.3.

1.2 The Configuration Space of Two Electrons

Consider the situation where two electrons with the same spin are lonesome in space. To simplify matters, place the origin at the center of mass. The Pauli exclusion principle dictates that the two electrons cannot be at the same place, so the electrons are somewhere outside the origin diametrically opposite of each other (assume they are point particles). However, you can't distinguish the two electrons, so the only thing you can tell is what line they are on, and how far they are from the origin (you can't give a vector v saying that this points at a chosen electron: $-v$ is just as good).

Disregarding the information telling you how far the electrons are from each other (which anyhow is just a matter of scale), we get that the space of possible positions may be identified with the *space of all lines through the origin* in $\mathbf{R}^3$. This space is called the (real) projective plane $\mathbf{RP}^2$. A line intersects the unit sphere $S^2 = \{p \in \mathbf{R}^3 \,|\, |p| = 1\}$ in exactly two (antipodal) points, and so we get that $\mathbf{RP}^2$ can be viewed as the sphere S^2 **but** with $p \in S^2$ identified with $-p$. A point in $\mathbf{RP}^2$ represented by $p \in S^2$ (and $-p$) is written $[p]$.

The projective plane is obviously a "manifold" (i.e., can be described by means of charts), since a neighborhood around $[p]$ can be identified with a neighborhood around $p \in S^2$ – as long as they are small enough to fit on one hemisphere. However, I cannot draw a picture of it in $\mathbf{R}^3$ without cheating.

On the other hand, there **is** a rather concrete representation of this space: it is what you get if you take a Möbius band (Figure 1.10) and a disk (Figure 1.11), and glue them together along their boundary (both the Möbius band and the disk have boundaries a copy of the circle). You are asked to perform this identification in Exercise 1.5.3.

1.2.1 Moral

The moral in this subsection is this: configuration spaces are oftentimes manifolds that do not in any natural way live in Euclidean space. From a technical point of view they often are what can be called *quotient spaces* (although this example was a rather innocent one in this respect).

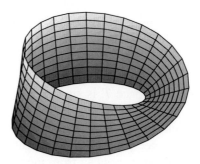

Figure 1.10. A Möbius band: note that its boundary is a circle.

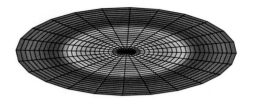

Figure 1.11. A disk: note that its boundary is a circle.

1.3 State Spaces and Fiber Bundles

The following example illustrates a phenomenon often encountered in physics, and a tool of vital importance for many applications. It is also an illustration of a key result which we will work our way towards: Ehresmann's fibration theorem, 8.5.10 (named after Charles Ehresmann, 1905–1979)[1].

It is slightly more involved than the previous example, since it points forward to many concepts and results we will discuss more deeply later, so if you find the going a bit rough, I advise you not to worry too much about the details right now, but come back to them when you are ready.

1.3.1 Qbits

In quantum computing one often talks about qbits. As opposed to an ordinary bit, which takes either the value 0 or the value 1 (representing "false" and "true" respectively), a *qbit*, or quantum bit, is represented by a complex linear combination ("superposition" in the physics parlance) of two states. The two possible states of a bit are then often called $|0\rangle$ and $|1\rangle$, and so a qbit is represented by the "pure qbit state" $\alpha|0\rangle + \beta|1\rangle$, where α and β are complex numbers and $|\alpha|^2 + |\beta|^2 = 1$ (since the total probability is 1, the numbers $|\alpha|^2$ and $|\beta|^2$ are interpreted as the probabilities that a measurement of the qbit will yield $|0\rangle$ and $|1\rangle$ respectively).

Note that the set of pairs $(\alpha, \beta) \in \mathbf{C}^2$ satisfying $|\alpha|^2 + |\beta|^2 = 1$ is just another description of the sphere $S^3 \subseteq \mathbf{R}^4 = \mathbf{C}^2$. In other words, a pure qbit state is a point (α, β) on the sphere S^3.

However, for various reasons *phase changes* are not important. A phase change is the result of multiplying $(\alpha, \beta) \in S^3$ by a unit-length complex number. That is, if $z = e^{i\theta} \in S^1 \subseteq \mathbf{C}$, the pure qbit state $(z\alpha, z\beta)$ is a phase shift of (α, β), and these should be identified. The *state space* is what you get when you identify each pure qbit state with the other pure qbit states you get by a phase change.

So, what is the relation between the space S^3 of pure qbit states and the state space? It turns out that the state space may be identified with the two-dimensional sphere S^2 (Figure 1.12), and the projection down to state space $\eta\colon S^3 \to S^2$ may then be given by

[1] https://en.wikipedia.org/wiki/Charles_Ehresmann

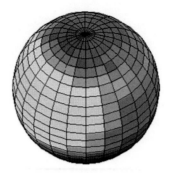

Figure 1.12. The state space S^2.

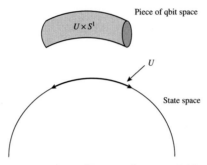

Figure 1.13. The pure qbit states represented in a small open neighborhood U in state space form a cylinder $U \times S^1$ (dimension reduced by one in the picture).

$$\eta(\alpha, \beta) = (|\alpha|^2 - |\beta|^2, 2\alpha\bar{\beta}) \in S^2 \subseteq \mathbf{R}^3 = \mathbf{R} \times \mathbf{C}.$$

Note that $\eta(\alpha, \beta) = \eta(z\alpha, z\beta)$ if $z \in S^1$, and so η does indeed send all the phase shifts of a given pure qbit to the same point in state space, and conversely, any two pure qbits in preimage of a given point in state space are phase shifts of each other.

Given a point in state space $p \in S^2$, the space of pure qbit states representing p can be identified with $S^1 \subseteq \mathbf{C}$: choose a pure qbit state (α, β) representing p, and note that any other pure qbit state representing p is of the form $(z\alpha, z\beta)$ for some *unique* $z \in S^1$.

So, can a pure qbit be given uniquely by its associated point in the state space and some point on the circle, i.e., is the space of pure qbit states really $S^2 \times S^1$ (and not S^3 as I previously claimed)? Without more work, it is not at all clear how these copies of S^1 lying over each point in S^2 are to be glued together: how does this "circle's worth" of pure qbit states change when we vary the position in state space slightly?

The answer comes through Ehresmann's fibration theorem, 8.5.10. It turns out that $\eta \colon S^3 \to S^2$ is a *locally* trivial fibration, which means that, in a small neighborhood U around any given point in state space, the space of pure qbit states *does* look like $U \times S^1$. See Figure 1.13. On the other hand, the *global* structure is different. In fact, $\eta \colon S^3 \to S^2$ is an important mathematical object for many reasons, and is known as the *Hopf fibration*.

The input to Ehresmann's theorem comes in two types. First we have some point set information, which in our case is handled by the fact that S^3 is "compact" A.7.1. Secondly, there is a condition which sees only the linear approximations, and which in our case boils down to the fact that any "infinitesimal" movement on S^2 is the shadow of an "infinitesimal" movement in S^3. This is a question which – given the right language – is settled through a quick and concrete calculation of differentials. We'll be more precise about this later (this is Exercise 8.5.16).

1.3.2 Moral

The idea is the important thing: if you want to understand some complicated model through some simplification, it is often so that the complicated model *locally* (in the simple model) can be built out of the simple model through multiplying with some fixed space.

How these local pictures are glued together to give the global picture is another matter, and often requires other tools, for instance from algebraic topology. In the $S^3 \to S^2$ case, we see that S^3 and $S^2 \times S^1$ cannot be identified since S^3 is simply connected (meaning that any closed loop in S^3 can be deformed continuously to a point) and $S^2 \times S^1$ is not.

An important class of examples (of which the above is one) of locally trivial fibrations arises from symmetries: if M is some (configuration) space and you have a "group of symmetries" G (e.g., rotations) acting on M, then you can consider the space M/G of points in M where you have identified two points in M if they can be obtained from each other by letting G act (e.g., one is a rotated copy of the other). Under favorable circumstances M/G will be a manifold and the projection $M \to M/G$ will be a locally trivial fibration, so that M is built by gluing together spaces of the form $U \times G$, where U varies over the open subsets of M/G.

1.4 Further Examples

A short bestiary of manifolds available to us at the moment might look like this.

- The surface of the earth, S^2, and higher-dimensional spheres, see Example 2.1.5.
- Space-time is a manifold: general relativity views space-time as a four-dimensional "pseudo-Riemannian" manifold. According to Einstein its curvature is determined by the mass distribution. (Whether the large-scale structure is flat or not is yet another question. Current measurements sadly seem to be consistent with a flat large-scale structure.)
- Configuration spaces in physics (e.g., the robot in Example 1.1, the two electrons of Example 1.2 or the more abstract considerations at the very end of Section 1.3.2 above).

- If $f : \mathbf{R}^n \to \mathbf{R}$ is a map and y a real number, then the inverse image

$$f^{-1}(y) = \{x \in \mathbf{R}^n \mid f(x) = y\}$$

  is often a manifold. For instance, if $f : \mathbf{R}^2 \to \mathbf{R}$ is the norm function $f(x) = |x|$, then $f^{-1}(1)$ is the unit circle S^1 (c.f. the discussion of submanifolds in Chapter 4).
- The torus (c.f. the robot in Example 1.1).
- "The real projective plane" $\mathbf{RP}^2 = \{$All lines in $\mathbf{R}^3$ through the origin$\}$ (see the two-electron case in Example 1.2, but also Exercise 1.5.3).
- The Klein bottle[2] (see Section 1.5).

We end this introduction by studying surfaces in a bit more detail (since they are concrete, and this drives home the familiar notion of charts in more exotic situations), and also come up with some inadequate words about higher-dimensional manifolds in general.

1.4.1 Charts

The space-time manifold brings home the fact that manifolds must be represented intrinsically: the surface of the earth is seen as a sphere "in space", but there is no space which should naturally harbor the universe, except the universe itself. This opens up the question of how one can determine the shape of the space in which we live.

One way of representing the surface of the earth as the two-dimensional space it is (not referring to some ambient three-dimensional space), is through an atlas. The shape of the earth's surface is then determined by how each map in the atlas is to be glued to the other maps in order to represent the entire surface.

Just like the surface of the earth is covered by maps, the torus in the robot's arm was viewed through flat representations. In the technical sense of the word, the representation was not a "chart" (see Definition 2.1.1) since some points were covered twice (just as Siberia and Alaska have a tendency to show up twice on some European maps). It is allowed to have many charts covering Fairbanks in our atlas, but on each single chart it should show up at most once. We may fix this problem at the cost of having to use more overlapping charts. Also, in the robot example (as well as the two-electron and qbit examples) we saw that it was advantageous to operate with more charts.

Example 1.4.2   To drive home this point, please play Jeff Weeks' "Torus Games" on www.geometrygames.org/TorusGames/ for a while.

....
[2] www-groups.dcs.st-and.ac.uk/~history/Biographies/Klein.html

1.5 Compact Surfaces

This section is rather autonomous, and may be read at leisure at a later stage to fill in the intuition on manifolds.

1.5.1 The Klein Bottle

To simplify we could imagine that we were two-dimensional beings living in a static closed surface. The sphere and the torus are familiar surfaces, but there are many more. If you did Example 1.4.2, you were exposed to another surface, namely the *Klein bottle*. This has a plane representation very similar to that of the torus: just reverse the orientation of a single edge (Figure 1.14).

Although the Klein bottle is an easy surface to describe (but frustrating to play chess on), it is too complicated to fit inside our three-dimensional space (again a manifold is **not** a space inside a flat space, it is a locally Euclidean space). The best we can do is to give an "immersed" (i.e., allowing self-intersections) picture (Figure 1.15).

Speaking of pictures: the Klein bottle makes a surprising entré in image analysis. When analyzing the nine-dimensional space of 3×3 patches of gray-scale pixels, it is of importance – for instance if you want to implement some compression technique – to know what high-contrast configurations occur most commonly. Carlsson, Ishkhanov, de Silva and Zomorodian show in [5] that the subspace of "most common high-contrast pixel configurations" actually "is" a Klein bottle.

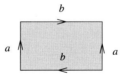

Figure 1.14. A plane representation of the Klein bottle: identify along the edges in the direction indicated.

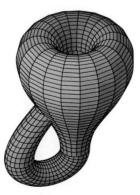

Figure 1.15. A picture of the Klein bottle forced into our three-dimensional space: it is really just a shadow since it has self-intersections. If you insist on putting this two-dimensional manifold into a flat space, you must have at least four dimensions available.

Their results have been used to develop a compression algorithm based on a "Klein bottle dictionary".

1.5.2 Classification of Compact Surfaces

As a matter of fact, it turns out that we can write down a list of all compact surfaces (*compact* is defined in Appendix A, but informally should be thought of as "closed and of bounded size"). First of all, surfaces may be divided into those that are orientable and those that are not. *Orientable* means that there are no loops by which two-dimensional beings living in the surface can travel and return home as their mirror images. (Is the universe non-orientable? Is that why some people are left-handed?)

All connected compact orientable surfaces can be obtained by attaching a finite number of handles to a sphere. The number of handles attached is referred to as the *genus* of the surface.

A *handle* is a torus with a small disk removed (see Figure 1.16). Note that the boundary of the hole on the sphere and the boundary of the hole on each handle are all circles, so we glue the surfaces together in a smooth manner along their common boundary (the result of such a gluing process is called the *connected sum*, and some care is required).

Thus all orientable compact surfaces are surfaces of pretzels with many holes (Figure 1.17).

There are nonorientable surfaces too (e.g., the Klein bottle). To make them, consider a Möbius band[3] (Figure 1.18). Its boundary is a circle, so after cutting a hole in a surface you may glue in a Möbius band. If you do this on a sphere you get the projective plane (this is Exercise 1.5.3). If you do it twice you get the Klein bottle. Any nonorientable compact surface can be obtained by cutting

Figure 1.16. A handle: ready to be attached to another 2-manifold with a small disk removed.

---

[3] www-groups.dcs.st-and.ac.uk/~history/Biographies/Mobius.html

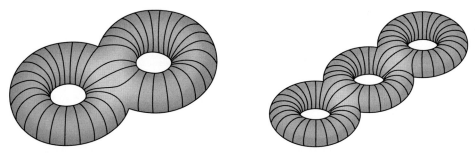

Figure 1.17. An orientable surface of genus *g* is obtained by gluing *g* handles (the smoothing out has yet to be performed in these pictures).

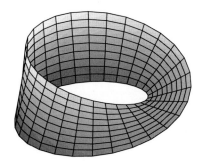

Figure 1.18. A Möbius band: note that its boundary is a circle.

a finite number of holes in a sphere and gluing in the corresponding number of Möbius bands.

The reader might wonder what happens if we mix handles and Möbius bands, and it is a strange fact that if you glue *g* handles and $h > 0$ Möbius bands you get the same as if you had glued $h + 2g$ Möbius bands! For instance, the projective plane with a handle attached is the same as the Klein bottle with a Möbius band glued onto it. But fortunately this is it; there are no more identifications among the surfaces.

So, any (connected compact) surface can be obtained by cutting *g* holes in S^2 and *either* gluing in *g* handles *or* gluing in *g* Möbius bands. For a detailed discussion the reader may turn to Chapter 9 of Hirsch's book [8].

1.5.3 Plane Models

If you find such descriptions elusive, you may derive some comfort from the fact that all compact surfaces can be described similarly to the way we described the torus. If we cut a hole in the torus we get a handle. This may be represented by plane models as in Figure 1.19: identify the edges as indicated.

If you want more handles you just glue many of these together, so that a *g*-holed torus can be represented by a 4*g*-gon where two and two edges are identified. (See Figure 1.20 for the case $g = 2$; the general case is similar. See

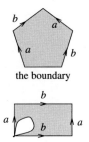

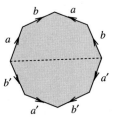

Figure 1.19. Two versions of a plane model for the handle: identify the edges as indicated to get a torus with a hole in.

Figure 1.20. A plane model of the orientable surface of genus two. Glue corresponding edges together. The dotted line splits the surface up into two handles.

Figure 1.21. A plane model for the Möbius band: identify the edges as indicated. When gluing it onto something else, use the boundary.

also www.rogmann.org/math/tori/torus2en.html for instruction on how to sew your own two- and three-holed torus.)

It is important to have in mind that the points on the edges in the plane models are in no way special: if we change our point of view slightly we can get them to be in the interior.

We have plane models for gluing in Möbius bands too (see Figure 1.21). So a surface obtained by gluing h Möbius bands to h holes on a sphere can be represented by a $2h$-gon, with pairwise identification of edges.

Example 1.5.1 If you glue two plane models of the Möbius band along their boundaries you get the picture in Figure 1.22. This represents the Klein bottle, but it is not exactly the same plane representation as the one we used earlier (Figure 1.14).

To see that the two plane models give the same surface, cut along the line c in the diagram on the left in Figure 1.23. Then take the two copies of the line a and glue them together in accordance with their orientations (this requires that you flip one

Figure 1.22. Gluing two flat Möbius bands together. The dotted line marks where the bands were glued together.

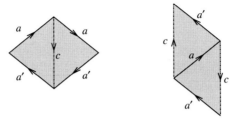

Figure 1.23. Cutting along c shows that two Möbius bands glued together amount to the Klein bottle.

of your triangles). The resulting diagram, which is shown to the right, is (a rotated and slanted version of) the plane model we used before for the Klein bottle.

Exercise 1.5.2 Prove by a direct cut-and-paste argument that what you get by adding a handle to the projective plane is the same as what you get if you add a Möbius band to the Klein bottle.

Exercise 1.5.3 Prove that the real projective plane

$$\mathbf{RP}^2 = \{\text{All lines in } \mathbf{R}^3 \text{ through the origin}\}$$

is the same as what you get by gluing a Möbius band to a sphere.

Exercise 1.5.4 See whether you can find out what the "Euler number"[4] (or Euler characteristic)[5] is. Then calculate it for various surfaces using the plane models. Can you see that both the torus and the Klein bottle have Euler number zero? The sphere has Euler number 2 (which leads to the famous theorem $V - E + F = 2$ for all surfaces bounding a "ball") and the projective plane has Euler number 1. The surface of Exercise 1.5.2 has Euler number -1. In general, adding a handle reduces the Euler number by two, and adding a Möbius band reduces it by one.

Exercise 1.5.5 If you did Exercise 1.5.4, design an (immensely expensive) experiment that could be performed by two-dimensional beings living in a compact orientable surface, determining the shape of their universe.

4 www-groups.dcs.st-and.ac.uk/~history/Biographies/Euler.html
5 http://en.wikipedia.org/wiki/Euler_characteristic

1.6 Higher Dimensions

Although surfaces are fun and concrete, next to no real-life applications are two-or three-dimensional. Usually there are zillions of variables at play, and so our manifolds will be correspondingly complex. This means that we can't continue to be vague (the previous sections indicated that even in three dimensions things become complicated). We need strict definitions to keep track of all the structure.

However, let it be mentioned at the informal level that we must not expect to have such a nice list of higher-dimensional manifolds as we had for compact surfaces. Classification problems for higher-dimensional manifolds constitute an extremely complex and interesting business we will not have occasion to delve into. It opens new fields of research using methods both from algebra and from analysis that go far beyond the ambitions of this text.

1.6.1 The Poincaré Conjecture and Thurston's Geometrization Conjecture

In 1904 H. Poincaré [6] conjectured that any simply connected compact and closed 3-manifold is homeomorphic to the 3-sphere. This problem remained open for almost 100 years, although the corresponding problem was resolved in higher dimensions by S. Smale [7] (1961; for dimensions greater than 4, see [18]) and M. Freedman [8] (1982; in dimension 4, see [7]).

In the academic year 2002/2003 G. Perelman [9] published a series of papers building on previous work by R. Hamilton [10], which by now have come to be widely regarded as the core of a proof of the Poincaré conjecture. The proof relies on an analysis of the "Ricci flow" deforming the curvature of a manifold in a manner somehow analogous to the heat equation, smoothing out irregularities. Our encounter with flows will be much more elementary, but will still prove essential in the proof of Ehresmann's fibration theorem, 8.5.10.

Perelman was offered the Fields Medal for his work in 2006, but spectacularly refused it. In this way he created much more publicity for the problem, mathematics and himself than would have otherwise been thinkable. In 2010 Perelman was also awarded the USD1M Millennium Prize from the Clay Mathematics Institute [11]. Again he turned down the prize, saying that Hamilton's contribution in proving the Poincaré conjecture was "no less than mine" (see, e.g., the Wikipedia entry [12] on the Poincaré conjecture for an updated account).

[6] www-groups.dcs.st-and.ac.uk/~history/Biographies/Poincare.html
[7] www-groups.dcs.st-and.ac.uk/~history/Biographies/Smale.html
[8] www-history.mcs.st-andrews.ac.uk/Mathematicians/Freedman.html
[9] http://en.wikipedia.org/wiki/Grigori_Perelman
[10] http://en.wikipedia.org/wiki/Richard_Hamilton
[11] www.claymath.org
[12] http://en.wikipedia.org/wiki/Poincare_conjecture

Of far greater consequence is Thurston's geometrization conjecture. This conjecture was proposed by W. Thurston[13] in 1982. Any 3-manifold can be decomposed into *prime manifolds*, and the conjecture says that any prime manifold can be cut along tori, so that the interior of each of the resulting manifolds has one of *eight geometric structures* with finite volume. See, e.g., the Wikipedia page[14] for further discussion and references to manuscripts with details of the proof filling in Perelman's sketch.

1.6.2 The History of Manifolds

Although it is a fairly young branch of mathematics, the history behind the theory of manifolds is rich and fascinating. The reader should take the opportunity to check out some of the biographies at The MacTutor History of Mathematics archive[15] or the Wikipedia entries of the mathematicians mentioned by name in the text (I have occasionally provided direct links).

There is also a page called History Topics: Geometry and Topology Index[16] which is worthwhile spending some time with. Of printed books, I have found Jean Dieudonné's book [6] especially helpful (although it is mainly concerned with topics beyond the scope of this book).

---

13 www-groups.dcs.st-and.ac.uk/~history/Biographies/Thurston.html
14 http://en.wikipedia.org/wiki/Geometrization_conjecture
15 www-groups.dcs.st-and.ac.uk/~history/index.html
16 www-groups.dcs.st-and.ac.uk/~history/Indexes/Geometry_Topology.html

2 Smooth Manifolds

2.1 Topological Manifolds

Let us get straight to our object of study. The terms used in the definition are explained immediately below the box. If words like "open" and "topology" are new to you, you are advised to read Appendix A on point-set topology in parallel with this chapter.

Definition 2.1.1 An *n-dimensional topological manifold* M is a Hausdorff topological space with a countable basis for the topology which is *locally homeomorphic* to $\mathbf{R}^n$.

The last point (*locally homeomorphic* to $\mathbf{R}^n$ – implicitly with the metric topology – also known as Euclidean space, see Definition A.1.8) means that for every point $p \in M$ there is

an open neighborhood U of p in M,
an open set $U' \subseteq \mathbf{R}^n$ and
a homeomorphism (Definition A.2.5) $\quad x : U \to U'$.

We call such an $x : U \to U'$ a *chart* and U a *chart domain* (Figure 2.1).

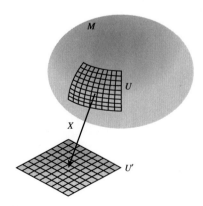

Figure 2.1.

A collection of charts $\{x_\alpha \colon U_\alpha \to U'_\alpha\}$ covering M (i.e., such that the union $\bigcup U_\alpha$ of the chart domains is M) is called an *atlas*.

Note 2.1.2 The conditions that M should be "Hausdorff" (Definition A.4.1) and have a "countable basis for its topology" (Section A.3) will not play an important rôle for us for quite a while. It is tempting to just skip these conditions, and come back to them later when they actually are important. As a matter of fact, on a first reading I suggest you actually do this. Rest assured that all subsets of Euclidean spaces satisfy these conditions (see Corollary A.5.6).

The conditions are there in order to exclude some pathological creatures that are locally homeomorphic to $\mathbf{R}^n$, but are so weird that we do not want to consider them. We include the conditions at once so as not to need to change our definition in the course of the book, and also to conform with usual language.

Note that $\mathbf{R}^n$ itself is a smooth manifold. In particular, it has a countable basis for its topology (c.f. Exercise A.3.4). The requirement that there should be a countable basis for the topology could be replaced by demanding the existence of a countable atlas.

Note 2.1.3 When saying that "M is a manifold" without specifying its dimension, one could envision that the dimension need not be the same everywhere. We only really care about manifolds of a fixed dimension, and even when allowing the dimension to vary, each connected component has a unique dimension. Consequently, you may find that we'll not worry about this and in the middle of an argument say something like "let n be the dimension of M" (and proceed to talk about things that concern only one component at a time).

Example 2.1.4 Let $U \subseteq \mathbf{R}^n$ be an open subset. Then U is an n-manifold. Its atlas needs only one chart, namely the identity map $\mathrm{id} \colon U = U$. As a sub-example we have the open n-disk

$$E^n = \{p \in \mathbf{R}^n \mid |p| < 1\}.$$

The notation E^n has its disadvantages. You may find it referred to as B^n, $B^n(1)$, $E^n(1)$, $B_1(0)$, $N_1^{\mathbf{R}^n}(0) \ldots$ in other texts.

Example 2.1.5 The n-sphere

$$S^n = \{p \in \mathbf{R}^{n+1} \mid |p| = 1\}$$

is an n-dimensional manifold. To see that S^n is locally homeomorphic to $\mathbf{R}^n$ we may proceed as follows. Write a point in $\mathbf{R}^{n+1}$ as an $n + 1$ tuple indexed from 0 to n: $p = (p_0, p_1, \ldots, p_n)$. To give an atlas for S^n, consider the open sets

$$U^{k,0} = \{p \in S^n \mid p_k > 0\},$$
$$U^{k,1} = \{p \in S^n \mid p_k < 0\}$$

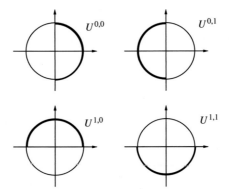

Figure 2.2.

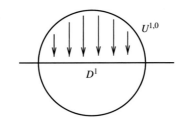

Figure 2.3.

for $k = 0, \ldots, n$, and let

$$x^{k,i} : U^{k,i} \to E^n$$

be the projection to the open n-disk E^n given by deleting the kth coordinate:

$$(p_0, \ldots, p_n) \mapsto (p_0, \ldots, \widehat{p_k}, \ldots, p_n)$$
$$= (p_0, \ldots, p_{k-1}, p_{k+1}, \ldots, p_n)$$

(the "hat" in $\widehat{p_k}$ is a common way to indicate that this coordinate should be deleted). See Figures 2.2 and 2.3.

(The n-sphere is Hausdorff and has a countable basis for its topology by Corollary A.5.6 simply because it is a subspace of $\mathbf{R}^{n+1}$.)

Exercise 2.1.6   Check that the proposed charts $x^{k,i}$ for S^n in the previous example really are homeomorphisms.

Exercise 2.1.7   We shall later see that an atlas with two charts suffices on the sphere. Why is there no atlas for S^n with only one chart?

Example 2.1.8   The real projective n-space $\mathbf{RP}^n$ is the set of all straight lines through the origin in $\mathbf{R}^{n+1}$. As a topological space, it is the quotient space (see Section A.6)

$$\mathbf{RP}^n = (\mathbf{R}^{n+1} \setminus \{0\})/\sim,$$

where the equivalence relation is given by $p \sim q$ if there is a nonzero real number λ such that $p = \lambda q$. Since each line through the origin intersects the unit sphere in two (antipodal) points, $\mathbf{RP}^n$ can alternatively be described as

$$S^n/\sim,$$

where the equivalence relation is $p \sim -p$. The real projective n-space is an n-dimensional manifold, as we shall see below. If $p = (p_0, \ldots, p_n) \in \mathbf{R}^{n+1} \setminus \{0\}$ we write $[p]$ for its equivalence class considered as a point in $\mathbf{RP}^n$.

For $0 \le k \le n$, let

$$U^k = \{[p] \in \mathbf{RP}^n | p_k \ne 0\}.$$

Varying k, this gives an open cover of $\mathbf{RP}^n$ (why is U^k open in $\mathbf{RP}^n$?). Note that the projection $S^n \to \mathbf{RP}^n$ when restricted to $U^{k,0} \cup U^{k,1} = \{p \in S^n | p_k \ne 0\}$ gives a two-to-one correspondence between $U^{k,0} \cup U^{k,1}$ and U^k. In fact, when restricted to $U^{k,0}$ the projection $S^n \to \mathbf{RP}^n$ yields a homeomorphism $U^{k,0} \cong U^k$.

The homeomorphism $U^{k,0} \cong U^k$ together with the homeomorphism

$$x^{k,0} \colon U^{k,0} \to E^n = \{p \in \mathbf{R}^n \mid |p| < 1\}$$

of Example 2.1.5 gives a chart $U^k \to E^n$ (the explicit formula is given by sending $[p] \in U^k$ to $(|p_k|/(p_k|p|)) (p_0, \ldots, \widehat{p_k}, \ldots, p_n)$. Letting k vary, we get an atlas for $\mathbf{RP}^n$.

We can simplify this somewhat: the following atlas will be referred to as the *standard atlas for* $\mathbf{RP}^n$. Let

$$x^k \colon U^k \to \mathbf{R}^n$$

$$[p] \mapsto \frac{1}{p_k} (p_0, \ldots, \widehat{p_k}, \ldots, p_n).$$

Note that this is well defined since

$$(1/p_k)(p_0, \ldots, \widehat{p_k}, \ldots, p_n) = (1/(\lambda p_k))(\lambda p_0, \ldots, \widehat{\lambda p_k}, \ldots, \lambda p_n).$$

Furthermore, x^k is a bijective function with inverse given by

$$\left(x^k\right)^{-1} (p_0, \ldots, \widehat{p_k}, \ldots, p_n) = [p_0, \ldots, 1, \ldots, p_n]$$

(note the convenient cheating in indexing the points in $\mathbf{R}^n$).

In fact, x^k is a homeomorphism: x^k is continuous since the composite $U^{k,0} \cong U^k \to \mathbf{R}^n$ is; and $\left(x^k\right)^{-1}$ is continuous since it is the composite $\mathbf{R}^n \to \{p \in \mathbf{R}^{n+1} \mid p_k \ne 0\} \to U^k$, where the first map is given by $(p_0, \ldots, \widehat{p_k}, \ldots, p_n) \mapsto (p_0, \ldots, 1, \ldots, p_n)$ and the second is the projection.

(That $\mathbf{RP}^n$ is Hausdorff and has a countable basis for its topology is shown in Exercise A.7.5.)

Note 2.1.9 It is not obvious at this point that $\mathbf{RP}^n$ can be realized as a subspace of a Euclidean space (we will show it can in Theorem 8.2.6).

Note 2.1.10 We will try to be consistent in letting the charts have names like x and y. This is sound practice since it reminds us that what charts are good for is to give "local coordinates" on our manifold: a point $p \in M$ corresponds to a point

$$x(p) = (x_1(p), \ldots, x_n(p)) \in \mathbf{R}^n.$$

The general philosophy when studying manifolds is to refer back to properties of Euclidean space by means of charts. In this manner a successful theory is built up: whenever a definition is needed, we take the Euclidean version and require that the corresponding property for manifolds is the one you get by saying that it must hold true in "local coordinates".

Example 2.1.11 As we defined it, a topological manifold is a topological space with certain properties. We could have gone about this differently, minimizing the rôle of the space at the expense of talking more about the atlas.

For instance, given a set M, a collection $\{U_\alpha\}_{\alpha \in A}$ of subsets of M such that $\bigcup_{\alpha \in A} U_\alpha = M$ (we say that $\{U_\alpha\}_{\alpha \in A}$ covers M) and a collection of injections (one-to-one functions) $\{x_\alpha \colon U_\alpha \to \mathbf{R}^n\}_{\alpha \in A}$, assume that if $\alpha, \beta \in A$ then the bijection $x_\alpha(U_\alpha \cap U_\beta) \to x_\beta(U_\alpha \cap U_\beta)$ sending q to $x_\beta x_\alpha^{-1}(q)$ is a continuous map between open subsets of $\mathbf{R}^n$.

The declaration that $U \subset M$ is open if for all $\alpha \in A$ we have that $x_\alpha(U \cap U_\alpha) \subseteq \mathbf{R}^n$ is open determines a topology on M. If this topology is Hausdorff and has a countable basis for its topology, then M is a topological manifold. This can be achieved if, for instance, we have that

(1) for $p, q \in M$, either there is an $\alpha \in A$ such that $p, q \in U_\alpha$ or there are $\alpha, \beta \in A$ such that U_α and U_β are disjoint with $p \in U_\alpha$ and $q \in U_\beta$ and
(2) there is a countable subset $B \subseteq A$ such that $\bigcup_{\beta \in B} U_\beta = M$.

2.2 Smooth Structures

We will have to wait until Definition 2.3.5 for the official definition of a *smooth manifold*. The idea is simple enough: in order to do *differential* topology we need that the charts of the manifolds are glued smoothly together, so that our questions regarding differentials or the like do not get different answers when interpreted through different charts. Again "smoothly" must be borrowed from the Euclidean world. We proceed to make this precise.

Let M be a topological manifold, and let $x_1 \colon U_1 \to U_1'$ and $x_2 \colon U_2 \to U_2'$ be two charts on M with U_1' and U_2' open subsets of $\mathbf{R}^n$. Assume that $U_{12} = U_1 \cap U_2$ is nonempty.

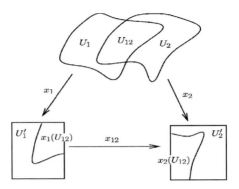

Figure 2.4.

Then we may define a *chart transformation* as shown in Figure 2.4 (sometimes called a "transition map")

$$x_{12} \colon x_1(U_{12}) \to x_2(U_{12})$$

by sending $q \in x_1(U_{12})$ to

$$x_{12}(q) = x_2 x_1^{-1}(q)$$

(in function notation we get that

$$x_{12} = \left(x_2|_{U_{12}}\right) \circ \left(x_1|_{U_{12}}\right)^{-1} \colon x_1(U_{12}) \to x_2(U_{12}),$$

where we recall that "$|_{U_{12}}$" means simply "restrict the domain of definition to U_{12}"). The picture of the chart transformation above will usually be recorded more succinctly as

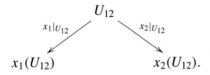

This makes things easier to remember than the occasionally awkward formulae. The restrictions, like in $x_1|_{U_{12}}$, clutter up the notation, and if we're pretty sure no confusion can arise we may in the future find ourselves writing variants like $x_2 x_1^{-1}|_{U_{12}}$ or even $x_2 x_1^{-1}$ when we should have written $(x_2|_{U_{12}})(x_1|_{U_{12}})^{-1}$. This is common practice, but in the beginning you should try to keep everything in place and relax your notation only once you are sure what you *actually* mean.

Definition 2.2.1 A map f between open subsets of Euclidean spaces is said to be *smooth* if all the higher-order partial derivatives exist and are continuous. A smooth map f between open subsets of $\mathbf{R}^n$ is said to be a *diffeomorphism* if it has a smooth inverse f^{-1}.

The chart transformation x_{12} is a function from an open subset of $\mathbf{R}^n$ to another, and it makes sense to ask whether it is smooth or not.

> **Definition 2.2.2**   An atlas on a manifold is *smooth* (or C^∞) if all the chart transformations are smooth.

Note 2.2.3   Note that, if x_{12} is a chart transformation associated with a pair of charts in an atlas, then x_{12}^{-1} is also a chart transformation. Hence, saying that an atlas is smooth is the same as saying that all the chart transformations are diffeomorphisms.

Note 2.2.4   We are interested only in the infinitely differentiable case, but in some situations it is sensible to ask for less. For instance, we could require that all chart transformations are C^1 (all the single partial differentials exist and are continuous). For a further discussion, see Note 2.3.7 below.

One could also ask for more, for instance that all chart transformations are analytic functions – giving the notion of an analytic manifold. However, the difference between smooth and analytic is substantial, as can be seen from Exercise 2.2.14.

Example 2.2.5   Let $U \subseteq \mathbf{R}^n$ be an open subset. Then the atlas whose only chart is the identity id: $U = U$ is smooth.

Example 2.2.6   The atlas

$$\mathcal{U} = \{(x^{k,i}, U^{k,i}) | 0 \le k \le n, 0 \le i \le 1\}$$

we gave on the n-sphere S^n in Example 2.1.5 is a smooth atlas. To see this, look at the example $U = U^{0,0} \cap U^{1,1}$ shown in Figure 2.5 and consider the associated chart transformation

$$\left(x^{1,1}|_U\right) \circ \left(x^{0,0}|_U\right)^{-1} : x^{0,0}(U) \to x^{1,1}(U).$$

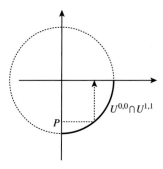

Figure 2.5. How the point p in $x^{0,0}(U)$ is mapped to $x^{1,1}(x^{0,0})^{-1}(p)$.

First we calculate the inverse of $x^{0,0}$: Let $p = (p_1, \ldots, p_n)$ be a point in the open disk E^n, then

$$\left(x^{0,0}\right)^{-1}(p) = \left(\sqrt{1 - |p|^2}, p_1, \ldots, p_n\right)$$

(we choose the positive square root, since we consider $x^{0,0}$). Furthermore,

$$x^{0,0}(U) = \{(p_1, \ldots, p_n) \in E^n \,|\, p_1 < 0\}.$$

Finally we get that if $p \in x^{0,0}(U)$ then

$$x^{1,1}\left(x^{0,0}\right)^{-1}(p) = \left(\sqrt{1 - |p|^2}, \widehat{p_1}, p_2, \ldots, p_n\right).$$

This is a smooth map, and on generalizing to other indices we get that we have a smooth atlas for S^n.

Example 2.2.7 There is another useful smooth atlas on S^n, given by *stereographic projection*. This atlas has only two charts, (x^+, U^+) and (x^-, U^-). The chart domains are

$$U^+ = \{p \in S^n \,|\, p_0 > -1\},$$
$$U^- = \{p \in S^n \,|\, p_0 < 1\},$$

and $x^+ : U^+ \to \mathbf{R}^n$ is given by sending a point p in S^n to the intersection $x^+(p)$ of the ("hyper") plane

$$\mathbf{R}^n = \{(0, p_1, \ldots, p_n) \in \mathbf{R}^{n+1}\}$$

and the straight line through the South pole $S = (-1, 0, \ldots, 0)$ and p (see Figure 2.6). Similarly for x^-, using the North pole instead. Note that both x^+ and x^- are homeomorphisms onto all of $\mathbf{R}^n$

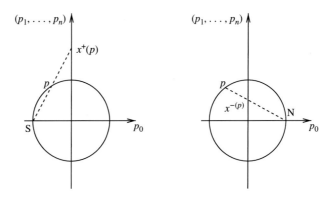

Figure 2.6.

To check that there are no unpleasant surprises, one should write down the formulae:

$$x^+(p) = \frac{1}{1 + p_0}(p_1, \ldots, p_n),$$
$$x^-(p) = \frac{1}{1 - p_0}(p_1, \ldots, p_n).$$

We observe that this defines homeomorphisms $U^\pm \cong \mathbf{R}^n$. We need to check that the chart transformations are smooth. Consider the chart transformation $x^+\left(x^-\right)^{-1}$ defined on $x^-(U^- \cap U^+) = \mathbf{R}^n \setminus \{0\}$. A small calculation gives that if $q \in \mathbf{R}^n$ then

$$\left(x^-\right)^{-1}(q) = \frac{1}{1 + |q|^2}(|q|^2 - 1, 2q)$$

(solve the equation $x^-(p) = q$ with respect to p), and so

$$x^+\left(x^-\right)^{-1}(q) = \frac{1}{|q|^2}q,$$

which is smooth. A similar calculation for the other chart transformation yields that $\{(x^+, U^+), (x^-, U^-)\}$ is a smooth atlas.

Exercise 2.2.8   Verify that the claims and formulae in the stereographic projection example are correct.

Note 2.2.9   The last two examples may be somewhat worrisome: the sphere is the sphere and these two atlases are two manifestations of the "same" sphere, are they not? We address questions of this kind in the next chapter, such as "When do two different atlases describe the same smooth manifold?" You should, however, be aware that there **are** "exotic" smooth structures on spheres, i.e., smooth atlases on the topological manifold S^n which describe smooth structures essentially different from the one(s?) we have described (but only in high dimensions). See in particular Exercise 2.3.10 and Note 2.4.13. Furthermore, there are topological manifolds which cannot be given smooth structures.

Example 2.2.10   The atlas we gave the real projective space is smooth. As an example consider the chart transformation $x^2(x^0)^{-1}$: if $p_2 \neq 0$ then

$$x^2\left(x^0\right)^{-1}(p_1, \ldots, p_n) = \frac{1}{p_2}(1, p_1, p_3, \ldots, p_n).$$

Exercise 2.2.11 Show in all detail that the complex projective n-space

$$\mathbf{CP}^n = (\mathbf{C}^{n+1} \setminus \{0\})/\sim,$$

where $z \sim w$ if there exists a $\lambda \in \mathbf{C} \setminus \{0\}$ such that $z = \lambda w$, is a compact $2n$-dimensional manifold. If your topology is not that strong yet, focus on the charts and chart transformations.

Exercise 2.2.12 There is a convenient smooth atlas for the circle S^1, whose charts we will refer to as *angle charts*. For each $\theta_0 \in \mathbf{R}$ consider the homeomorphism

$$(\theta_0, \theta_0 + 2\pi) \to S^1 - \{e^{i\theta_0}\}$$

given by sending θ to $e^{i\theta}$. Call the inverse x_{θ_0}. Check that $\{(x_{\theta_0}, S^1 - e^{i\theta_0})\}_{\theta_0}$ is a smooth atlas.

Exercise 2.2.13 Give the boundary of the square the structure of a smooth manifold.

Exercise 2.2.14 Let $\lambda : \mathbf{R} \to \mathbf{R}$ be defined by

$$\lambda(t) = \begin{cases} 0 & \text{for } t \leq 0 \\ e^{-1/t} & \text{for } t > 0. \end{cases}$$

This is a smooth function with values between zero and one. Note that all derivatives at zero are zero: the McLaurin series fails miserably and λ is definitely not analytic.

2.3 Maximal Atlases

We easily see that some manifolds can be equipped with many different smooth atlases. An example is the circle. Stereographic projection gives a different atlas than what you get if you for instance parametrize by means of the angle (Example 2.2.7 vs. Exercise 2.2.12). But we do not want to distinguish between these two "smooth structures", and in order to systematize this we introduce the concept of a *maximal atlas*.

Definition 2.3.1 Let M be a manifold and $\mathcal{A}$ a smooth atlas on M. Then we define $\mathcal{D}(\mathcal{A})$ as the following set of charts on M:

$$\mathcal{D}(\mathcal{A}) = \left\{ \text{charts } y \colon V \to V' \text{ on } M \;\middle|\; \begin{array}{l} \text{for all charts } (x, U) \text{ in } \mathcal{A}, \text{ the composite} \\ \quad x|_W (y|_W)^{-1} \colon y(W) \to x(W) \\ \text{is a diffeomorphism, where } W = U \cap V \end{array} \right\}.$$

Lemma 2.3.2 *Let M be a manifold and $\mathcal{A}$ a smooth atlas on M. Then $\mathcal{D}(\mathcal{A})$ is a smooth atlas.*

Proof. Let $y\colon V \to V'$ and $z\colon W \to W'$ be two charts in $\mathcal{D}(\mathcal{A})$. We have to show that

$$z|_{V\cap W} \circ (y|_{V\cap W})^{-1}$$

is smooth. Let q be any point in $y(V \cap W)$. We prove that $z \circ y^{-1}$ is smooth in a neighborhood of q. Choose a chart $x\colon U \to U'$ in $\mathcal{A}$ with $y^{-1}(q) \in U$.

Letting $O = U \cap V \cap W$, we get that

$$z|_O \circ (y|_O)^{-1} = z|_O \circ ((x|_O)^{-1} \circ x|_O) \circ (y|_O)^{-1}$$
$$= \left(z|_O \circ (x|_O)^{-1}\right) \circ \left(x|_O \circ (y|_O)^{-1}\right).$$

Since y and z are in $\mathcal{D}(\mathcal{A})$ and x is in $\mathcal{A}$ we have by definition that both the maps in the composite above are smooth, and we are done. ∎

The crucial equation can be visualized by the following diagram:

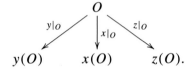

$$y(O) \qquad x(O) \qquad z(O).$$

Going up and down with $x|_O$ in the middle leaves everything fixed so the two functions from $y(O)$ to $z(O)$ are equal.

Definition 2.3.3 A smooth atlas is *maximal* if there is no strictly bigger smooth atlas containing it.

Exercise 2.3.4 Given a smooth atlas $\mathcal{A}$, prove that $\mathcal{D}(\mathcal{A})$ is maximal. Hence any smooth atlas is a subset of a unique maximal smooth atlas.

> **Definition 2.3.5** A *smooth structure* on a topological manifold is a maximal smooth atlas. A *smooth manifold* $(M, \mathcal{A})$ is a topological manifold M equipped with a smooth structure $\mathcal{A}$. A *differentiable manifold* is a topological manifold for which there exists a smooth structure.

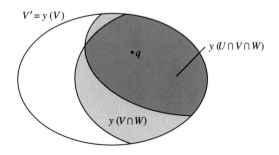

Figure 2.7.

Note 2.3.6   The following terms are synonymous: smooth, differential and C^∞.

Note 2.3.7   We are interested only in the smooth case, but in some situations it is sensible to ask for less. For instance, we could require that all chart transformations are C^1 (all the single partial differentials exist and are continuous). However, the distinction is not really important since having an atlas with C^1 chart transformations implies that there is a unique maximal smooth atlas such that the mixed chart transformations are C^1 (see, e.g., Theorem 2.9 in Chapter 2 of [8]).

Note 2.3.8   In practice we do not give the maximal atlas, but choose only a small practical smooth atlas and apply $\mathcal{D}$ to it. Often we write just M instead of $(M, \mathcal{A})$ if $\mathcal{A}$ is clear from the context.

Exercise 2.3.9	To check that two smooth atlases $\mathcal{A}$ and $\mathcal{B}$ give the same smooth structure on M (i.e., that $\mathcal{D}(\mathcal{A}) = \mathcal{D}(\mathcal{B})$) it suffices to verify that for each $p \in M$ there are charts $(x, U) \in \mathcal{A}$ and $(y, V) \in \mathcal{B}$ with $p \in W = U \cap V$ such that $x|_W (y|_W)^{-1} \colon y(W) \to x(W)$ is a diffeomorphism.
Exercise 2.3.10	Show that the two smooth atlases we have defined on S^n (the standard atlas in Example 2.1.5 and the stereographic projections of Example 2.2.7) are contained in a common maximal atlas. Hence they define the same smooth manifold, which we will simply call the *(standard smooth) sphere*.
Exercise 2.3.11	Choose your favorite diffeomorphism $x \colon \mathbf{R}^n \to \mathbf{R}^n$. Why is the smooth structure generated by x equal to the smooth structure generated by the identity? What does the maximal atlas for this smooth structure (the only one we'll ever consider) on $\mathbf{R}^n$ look like?
Exercise 2.3.12	Prove that any smooth manifold $(M, \mathcal{A})$ has a countable smooth atlas $\mathcal{V}$ (so that $\mathcal{D}(\mathcal{V}) = \mathcal{A}$).
Exercise 2.3.13	Prove that the atlas given by the angle charts in Exercise 2.2.12 gives the standard smooth structure on S^1.

Following up Example 2.1.11 we see that we can construct smooth manifolds from scratch, without worrying too much about the topology.

Lemma 2.3.14   *Given*

(1) a set M,
(2) a collection $\mathcal{A}$ of subsets of M, and
(3) an injection $x_U \colon U \to \mathbf{R}^n$ for each $U \in \mathcal{A}$,

such that

(1) there is a countable subcollection of $\mathcal{A}$ which covers M,

(2) for $p, q \in M$, either there is a $U \in \mathcal{A}$ such that $p, q \in U$ or there are $U, V \in \mathcal{A}$ such that U and V are disjoint with $p \in U$ and $q \in V$, and

(3) if $U, V \in \mathcal{A}$ then the bijection $x_U(U \cap V) \to x_V(U \cap V)$ sending q to $x_V x_U^{-1}(q)$ is a smooth map between open subsets of $\mathbf{R}^n$,

then there is a unique topology on M such that $(M, \mathcal{D}(\{(x_U, U)\}_{U \in \mathcal{A}}))$ is a smooth manifold.

Proof. For the x_Us to be homeomorphisms we *must* have that a subset $W \subseteq M$ is open if and only if for all $U \in \mathcal{A}$ the set $x_U(U \cap W)$ is an open subset of $\mathbf{R}^n$. As before, M is a topological manifold, and by the last condition $\{(x_U, U)\}_{U \in \mathcal{A}}$ is a smooth atlas. ∎

Example 2.3.15 As an example of how Lemma 2.3.14 can be used to construct smooth manifolds, we define a family of very important smooth manifolds called the Grassmann manifolds (after Hermann Grassmann (1809–1877))[1]. These manifolds show up in a number of applications, and are important to the theory of vector bundles (see for instance Section 6.8).

For $0 < k \leq n$, let

$$\mathrm{Gr}(k, \mathbf{R}^n)$$

(the notation varies in the literature) be the set of all k-dimensional linear subspaces of $\mathbf{R}^n$. Note that $\mathrm{Gr}(1, \mathbf{R}^{n+1})$ is nothing but the projective space $\mathbf{RP}^n$.

We will equip $\mathrm{Gr}(k, \mathbf{R}^n)$ with the structure of an $(n-k)k$-dimensional smooth manifold, the *Grassmann manifold*.

Doing this properly requires some care, but it is worth your while in that it serves the dual purpose of making the structure of this important space clearer as well as driving home some messages about projections that your linear algebra course might have been too preoccupied with multiplying matrices to make apparent.

If $V, W \subseteq \mathbf{R}^n$ are linear subspaces, we let $\mathrm{pr}^V : \mathbf{R}^n \to V$ be the orthogonal projection to V (with the usual inner product) and $\mathrm{pr}^V_W : W \to V$ the restriction of pr^V to W.

We let $\mathrm{Hom}(V, W)$ be the vector space of all linear maps from V to W. Concretely, and for the sake of the smoothness arguments below, using the standard basis for $\mathbf{R}^n$ we may identify $\mathrm{Hom}(V, W)$ with the $\dim(V) \cdot \dim(W)$-dimensional linear subspace of the space of $n \times n$ matrices A with the property that, if $v \in V$ and $v' \in V^\perp$, then $Av \in W$ and $Av' = 0$. (Check that the map

$$\{A \in M_n(\mathbf{R}) \mid v \in V, v' \in V^\perp \Rightarrow Av \in W, Av' = 0\} \to \mathrm{Hom}(V, W)$$
$$A \mapsto \{v \mapsto Av\}$$

----

[1] https://en.wikipedia.org/wiki/Hermann_Grassmann

is an isomorphism with inverse given by representing a linear transformation as a matrix.)

If $V \in \mathrm{Gr}(k, \mathbf{R}^n)$, consider the set

$$U_V = \{W \in \mathrm{Gr}(k, \mathbf{R}^n) \mid W \cap V^\perp = 0\}.$$

We will give "charts" of the form

$$x_V : U_V \to \mathrm{Hom}(V, V^\perp),$$

to which the reader might object that $\mathrm{Hom}(V, V^\perp)$ is not as such an open subset of Euclidean space. This is not of the essence, because $\mathrm{Hom}(V, V^\perp)$ is isomorphic to the vector space $M_{(n-k)k}(\mathbf{R})$ of all $(n-k) \times k$ matrices (choose bases $V \cong \mathbf{R}^k$ and $V^\perp \cong \mathbf{R}^{n-k}$), which again is isomorphic to $\mathbf{R}^{(n-k)k}$.

Another characterization of U_V is as the set of all $W \in \mathrm{Gr}(k, \mathbf{R}^n)$ such that $\mathrm{pr}_W^V : W \to V$ is an isomorphism. Let $x_V : U_V \to \mathrm{Hom}(V, V^\perp)$ send $W \in U_V$ to the composite

$$x_V(W) : V \xrightarrow{(\mathrm{pr}_W^V)^{-1}} W \xrightarrow{\mathrm{pr}_W^{V^\perp}} V^\perp.$$

See Figure 2.8. Varying V we get a smooth atlas $\{(x_V, U_V)\}_{V \in \mathrm{Gr}(k, \mathbf{R}^n)}$ for $\mathrm{Gr}(k, \mathbf{R}^n)$.

Exercise 2.3.16 Prove that the proposed structure on the Grassmann manifold $\mathrm{Gr}(k, \mathbf{R}^n)$ in Example 2.3.15 actually is a smooth atlas which endows $\mathrm{Gr}(k, \mathbf{R}^n)$ with the structure of a smooth manifold.

Note 2.3.17 In Chapter 1 we used the word "boundary" for familiar objects like the closed unit disk (whose boundary is the unit circle). Generally, the notion of a *smooth n-dimensional manifold with boundary M* is defined exactly as we defined a smooth *n*-dimensional manifold, except that the charts $x : U \to U'$ in our atlas are required to be homeomorphisms to open subsets U' of the *half space*

$$\mathbf{H}^n = \{(p_1, \ldots, p_n) \in \mathbf{R}^n \mid p_1 \geq 0\}$$

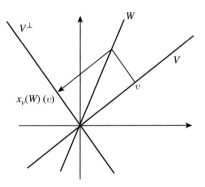

Figure 2.8.

(i.e., $U' = \mathbf{H}^n \cap \tilde{U}'$, where $\tilde{U}'$ is an open subset of $\mathbf{R}^n$). If $y\colon V \to V'$ is another chart in the atlas, the chart transformation $y \circ x^{-1}|_{x(U \cap V)}$ is "smooth", but what do we mean by a smooth map $f\colon V' \to W'$ when V' and W' are open subsets of half spaces? Here is the general definition: let $f\colon V' \to W'$, where $V' \subseteq \mathbf{R}^m$ and $W' \subseteq \mathbf{R}^n$ are arbitrary subsets. We say that f is *smooth* if for each $p \in V'$ there exist an open neighborhood $p \in \tilde{V}' \subseteq \mathbf{R}^m$ and a smooth map $\tilde{f}\colon \tilde{V}' \to \mathbf{R}^n$ such that for each $q \in V' \cap \tilde{V}'$ we have that $f(q) = \tilde{f}(q)$.

If M is a manifold with boundary, then its *boundary* ∂M is the subspace of points mapped to the boundary $\partial \mathbf{H} = \{(0, p_2, \dots, p_n)\}$ by some (and hence, it turns out, all) charts. The boundary is an $(n-1)$-dimensional manifold (without boundary). As an example, consider the closed n-dimensional unit ball; it is an n-dimensional manifold with boundary the unit $(n-1)$-sphere. Ordinary smooth manifolds correspond to manifolds with empty boundary.

2.4 Smooth Maps

Having defined smooth manifolds, we need to define smooth maps between them. No surprise: smoothness is a local question, so we may fetch the notion from Euclidean space by means of charts. (See Figure 2.9.)

> **Definition 2.4.1** Let $(M, \mathcal{A})$ and $(N, \mathcal{B})$ be smooth manifolds and $p \in M$. A continuous map $f\colon M \to N$ is *smooth at p* (or *differentiable at p*) if for any chart $x\colon U \to U' \in \mathcal{A}$ with $p \in U$ and any chart $y\colon V \to V' \in \mathcal{B}$ with $f(p) \in V$ the map
> $$y \circ f|_{U \cap f^{-1}(V)} \circ (x|_{U \cap f^{-1}(V)})^{-1}\colon x(U \cap f^{-1}(V)) \to V'$$
> is smooth at $x(p)$.
> We say that f is a *smooth map* if it is smooth at all points of M.

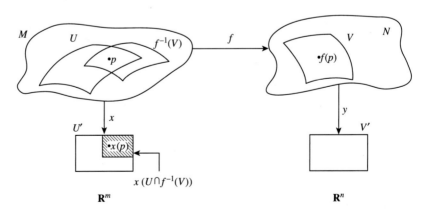

Figure 2.9.

Figure 2.9 will often find a less typographically challenging expression: "go up, over and down in the picture

$$W \xrightarrow{\; f|_W \;} V$$
$$x|_W \downarrow \qquad \qquad y \downarrow$$
$$x(W) \qquad \quad V',$$

where $W = U \cap f^{-1}(V)$, and see whether you have a smooth map of open subsets of Euclidean spaces". To see whether f in Definition 2.4.1 is smooth at $p \in M$ you do not actually have to check **all** charts.

Lemma 2.4.2 *Let $(M, \mathcal{A})$ and $(N, \mathcal{B})$ be smooth manifolds. A function $f : M \to N$ is smooth if (and only if) for all $p \in M$ there exist charts $(x, U) \in \mathcal{A}$ and $(y, V) \in \mathcal{B}$ with $p \in W = U \cap f^{-1}(V)$ such that the composite*

$$y \circ f|_W \circ (x|_W)^{-1} : x(W) \to y(V)$$

is smooth.

Exercise 2.4.3 Prove Lemma 2.4.2 (a full solution is provided in Appendix B, but you should really try yourself).

Exercise 2.4.4 Show that the map $\mathbf{R} \to S^1$ sending $p \in \mathbf{R}$ to $e^{ip} = (\cos p, \sin p) \in S^1$ is smooth.

Exercise 2.4.5 Show that the map $g : \mathbf{S}^2 \to \mathbf{R}^4$ given by

$$g(p_0, p_1, p_2) = (p_1 p_2, p_0 p_2, p_0 p_1, p_0^2 + 2p_1^2 + 3p_2^2)$$

defines a smooth injective map

$$\tilde{g} : \mathbf{RP}^2 \to \mathbf{R}^4$$

via the formula $\tilde{g}([p]) = g(p)$ (remember that $|p| = 1$; if you allow $p \in \mathbf{R}^3 - \{0\}$, you should use $g(p/|p|)$).

Exercise 2.4.6 Show that a map $f : \mathbf{RP}^n \to M$ is smooth if and only if the composite

$$S^n \xrightarrow{g} \mathbf{RP}^n \xrightarrow{f} M$$

is smooth, where g is the projection.

Definition 2.4.7 A smooth map $f : M \to N$ is a *diffeomorphism* if it is a bijection, and the inverse is smooth too. Two smooth manifolds are *diffeomorphic* if there exists a diffeomorphism between them.

Note 2.4.8 Note that this use of the word diffeomorphism coincides with the one used earlier for open subsets of $\mathbf{R}^n$.

Example 2.4.9 The smooth map $\mathbf{R} \to \mathbf{R}$ sending $p \in \mathbf{R}$ to p^3 is a smooth homeomorphism, but it is not a diffeomorphism: the inverse is not smooth at $0 \in \mathbf{R}$. The problem is that the derivative is zero at $0 \in \mathbf{R}$: if a smooth bijective map $f \colon \mathbf{R} \to \mathbf{R}$ has a nowhere-vanishing derivative, then it is a diffeomorphism. The inverse function theorem, 4.2.1, gives the corresponding criterion for (local) smooth invertibility also in higher dimensions.

Example 2.4.10 If $a < b \in \mathbf{R}$, then the straight line $f(t) = (b-a)t + a$ gives a diffeomorphism $f \colon (0,1) \to (a,b)$ with inverse given by $f^{-1}(t) = (t-a)/(b-a)$. Note that

$$\tan \colon (-\pi/2, \pi/2) \to \mathbf{R}$$

is a diffeomorphism. Hence all open intervals are diffeomorphic to the entire real line.

Exercise 2.4.11 Show that $\mathbf{RP}^1$ and S^1 are diffeomorphic.

Exercise 2.4.12 Show that $\mathbf{CP}^1$ and S^2 are diffeomorphic.

Note 2.4.13 The distinction between differentiable and smooth of Definition 2.3.5 (i.e., whether there merely exists a smooth structure or one has been chosen) is not always relevant, but the reader may find pleasure in knowing that according to Kervaire and Milnor [10] the topological manifold S^7 has 28 different smooth structures (up to "oriented" diffeomorphism, see Section 6.7 – 15 if orientation is ignored), and $\mathbf{R}^4$ has uncountably many [21].

As a side remark, one should notice that most physical situations involve differential equations of some sort, and so depend on the smooth structure, and not only on the underlying topological manifold. For instance, Baez remarks in This Week's Finds in Mathematical Physics *(Week 141)*, see `www.classe.cornell.edu/spr/1999-12/msg0019934.html`, that all of the 992 smooth structures on the 11-sphere are relevant to string theory.

Once one accepts the idea that there may be many smooth structures, one starts wondering what manifolds have a *unique* smooth structure (up to diffeomorphism). An amazing result in this direction recently appeared: Wang and Xu [22] have proved that the only odd-dimensional spheres with a unique smooth structure are S^1, S^3, S^5 and S^{61}(!). The even-dimensional case is not fully resolved; S^4 is totally mysterious, but apart from that one knows that S^2, S^6 and S^{56} are the only even-dimensional spheres in a range of dimensions that support exactly one smooth structure (at the time of writing it has been checked by Behrens, Hill, Hopkins and Ravenel, building on computations by Isaksen and Xu, up to dimension 140).

> **Lemma 2.4.14** *If* $f\colon (M,\mathcal{U}) \to (N,\mathcal{V})$ *and* $g\colon (N,\mathcal{V}) \to (P,\mathcal{W})$ *are smooth, then the composite* $gf\colon (M,\mathcal{U}) \to (P,\mathcal{W})$ *is smooth too.*

Proof. This is true for maps between Euclidean spaces, and we lift this fact to smooth manifolds. Let $p \in M$ and choose appropriate charts

$x\colon U \to U' \in \mathcal{U}$, such that $p \in U$,
$y\colon V \to V' \in \mathcal{V}$, such that $f(p) \in V$,
$z\colon W \to W' \in \mathcal{W}$, such that $gf(p) \in W$.

Then $T = U \cap f^{-1}(V \cap g^{-1}(W))$ is an open set containing p, and we have that

$$zgfx^{-1}|_{x(T)} = (zgy^{-1})(yfx^{-1})|_{x(T)},$$

which is a composite of smooth maps of Euclidean spaces, and hence smooth. ∎

In a picture, if $S = V \cap g^{-1}(W)$ and $T = U \cap f^{-1}(S)$:

$$
\begin{array}{ccccc}
T & \xrightarrow{\;f|_T\;} & S & \xrightarrow{\;g|_S\;} & W \\
\downarrow{\scriptstyle x|_T} & & \downarrow{\scriptstyle y|_S} & & \downarrow{\scriptstyle z|_W} \\
x(T) & & y(S) & & z(W).
\end{array}
$$

Going up and down with y does not matter.

Exercise 2.4.15 Let $f\colon M \to X$ be a homeomorphism of topological spaces. If M is a smooth manifold then there is a unique smooth structure on X that makes f a diffeomorphism.

In particular, note that, if $M = X = \mathbf{R}$ and $f\colon M \to X$ is the homeomorphism given by $f(t) = t^3$, then the above gives a *new* smooth structure on $\mathbf{R}$, but now (with respect to this structure) $f\colon M \to X$ is a diffeomorphism (as opposed to what was the case in Example 2.4.9), so the two smooth manifolds are diffeomorphic.

Definition 2.4.16 Let $(M,\mathcal{U})$ and $(N,\mathcal{V})$ be smooth manifolds. Then we let

$$\mathcal{C}^\infty(M, N) = \{\text{smooth maps } M \to N\}$$

and

$$\mathcal{C}^\infty(M) = \mathcal{C}^\infty(M, \mathbf{R}).$$

Note 2.4.17 A small digression, which may be disregarded if it contains words you haven't heard before. The outcome of the discussion above is that we have a category $\mathcal{C}^\infty$ of smooth manifolds: the objects are the smooth manifolds, and, if M and N are smooth, then

$$\mathcal{C}^\infty(M, N)$$

is the set of morphisms. The statement that $\mathcal{C}^\infty$ is a category uses that the identity map is smooth (check), and that the composition of smooth functions is smooth, giving the composition in $\mathcal{C}^\infty$:

$$\mathcal{C}^\infty(N, P) \times \mathcal{C}^\infty(M, N) \to \mathcal{C}^\infty(M, P).$$

The diffeomorphisms are the isomorphisms in this category.

Definition 2.4.18 A smooth map $f : M \to N$ is a *local diffeomorphism* if for each $p \in M$ there is an open set $U \subseteq M$ containing p such that $f(U)$ is an open subset of N and

$$f|_U : U \to f(U)$$

is a diffeomorphism.

Example 2.4.19 The projection $S^n \to \mathbf{RP}^n$ is a local diffeomorphism (Figure 2.10). Here is a more general example: let M be a smooth manifold, and

$$i : M \to M$$

a diffeomorphism with the property that $i(p) \neq p$, but $i(i(p)) = p$ for all $p \in M$ (such an animal is called a *fixed point free involution*). The quotient space M/i gotten by identifying p and $i(p)$ has a smooth structure, such that the projection

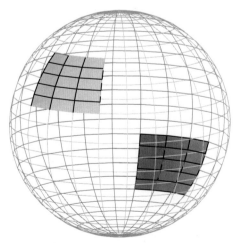

Figure 2.10. Small open sets in $\mathbf{RP}^2$ correspond to unions $U \cup (-U)$, where $U \subseteq S^2$ is an open set totally contained in one hemisphere.

$f \colon M \to M/i$ is a local diffeomorphism. We leave the proof of this claim as an exercise.

Exercise 2.4.20 Show that M/i has a smooth structure such that the projection $f \colon M \to M/i$ is a local diffeomorphism.

Exercise 2.4.21 If $(M, \mathcal{U})$ is a smooth n-dimensional manifold and $p \in M$, then there is a chart $x \colon U \to \mathbf{R}^n$ such that $x(p) = 0$.

Note 2.4.22 In differential topology one considers two smooth manifolds to be the same if they are diffeomorphic, and all properties one studies are unaffected by diffeomorphisms.

Is it possible to give a classification of manifolds? That is, can we list all the smooth manifolds? On the face of it this is a totally over-ambitious question, but actually quite a lot is known, especially about the compact (Definition A.7.1) connected (Definition A.9.1) smooth manifolds.

The circle is the only compact connected smooth 1-manifold.

In dimension two it is only slightly more interesting. As we discussed in Section 1.5, you can obtain any compact (smooth) connected 2-manifold by punching g holes in the sphere S^2 and glue onto this either g handles or g Möbius bands.

In dimension four and up total chaos reigns (and so it is here that most of the interesting stuff is to be found). Well, actually only the part within the parentheses is true in the last sentence: there is a lot of structure, much of it well understood. However, all of it is beyond the scope of this text. It involves quite a lot of manifold theory, but also algebraic topology and a subject called surgery, which in spirit is not so distant from the cutting and pasting techniques we used on surfaces in Section 1.5. For dimension three, the reader may refer back to Section 1.6.1.

Note 2.4.23 The notion of a smooth map between manifolds with boundary is defined exactly as for ordinary manifolds, except that we need to use the extension of the notion of a smooth map $V' \to W'$ to cover the case where V' and W' are open subsets of half spaces as explained in Note 2.3.17.

2.5 Submanifolds

What should a smooth submanifold be? Well, for sure,

$$\mathbf{R}^n \times \{0\} = \{(t_1, \ldots, t_{n+k}) \in \mathbf{R}^{n+k} \mid t_{n+1} = \cdots = t_{n+k} = 0\}$$

ought to be a smooth submanifold of $\mathbf{R}^n \times \mathbf{R}^k = \mathbf{R}^{n+k}$, and – just as we modeled smooth manifolds locally by Euclidean space – we use this example to model the concept of a smooth submanifold (see Figure 2.11). At this point it is perhaps not entirely clear that this will cover all the examples we are interested in. However, we will see somewhat later (more precisely, in Theorem 4.7.4) that this definition is

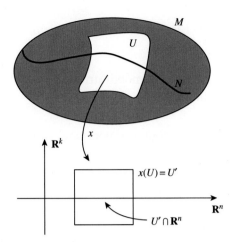

Figure 2.11.

equivalent to another, more conceptual and effectively checkable definition, which we as yet do not have all the machinery to formulate. Regardless, submanifolds are too important for us to afford to wait.

Definition 2.5.1   Let $(M, \mathcal{U})$ be a smooth $(n + k)$-dimensional manifold.

An n-dimensional *smooth submanifold* in M (Figure 2.11) is a subset $N \subseteq M$ such that for each $p \in N$ there is a chart $x \colon U \to U'$ in $\mathcal{U}$ with $p \in U$ such that

$$x(U \cap N) = U' \cap (\mathbf{R}^n \times \{0\}) \subseteq \mathbf{R}^n \times \mathbf{R}^k.$$

We say that the *codimension* of N in M is k.

In this definition we identify $\mathbf{R}^{n+k}$ with $\mathbf{R}^n \times \mathbf{R}^k$. We often write $\mathbf{R}^n \subseteq \mathbf{R}^n \times \mathbf{R}^k$ instead of $\mathbf{R}^n \times \{0\} \subseteq \mathbf{R}^n \times \mathbf{R}^k$ to signify the subset of all points with the last k coordinates equal to zero.

Note 2.5.2   The language of the definition really makes some sense: if $(M, \mathcal{U})$ is a smooth manifold and $N \subseteq M$ a submanifold, then N inherits a smooth structure such that the inclusion $N \to M$ is smooth. If $p \in N$ choose a chart (x_p, U_p) on M with $p \in U_p$ such that $x_p(U_p \cap N) = x_p(U_p) \cap (\mathbf{R}^n \times \{0\})$. Restricting to $U_p \cap N$ and projecting to the first n coordinates gives a homeomorphism from $U_p \cap N$ to an open subset of $\mathbf{R}^n$. On letting p vary we get a smooth atlas for N (the chart transformations consist of restrictions of chart transformations in $\mathcal{U}$).

Example 2.5.3   Let n be a natural number. Then $K_n = \{(p, p^n)\} \subseteq \mathbf{R}^2$ is a smooth submanifold. We define a smooth chart

$$x \colon \mathbf{R}^2 \to \mathbf{R}^2, \qquad (p, q) \mapsto (p, q - p^n).$$

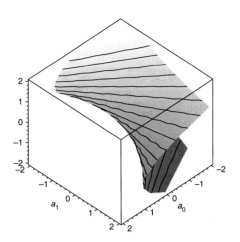

Figure 2.12.

Note that as required, x is smooth, with smooth inverse given by

$$(p, q) \mapsto (p, q + p^n),$$

and that $x(K_n) = \mathbf{R}^1 \times \{0\}$.

Exercise 2.5.4 Prove that $S^1 \subset \mathbf{R}^2$ is a submanifold. More generally, prove that $S^n \subset \mathbf{R}^{n+1}$ is a submanifold.

Exercise 2.5.5 Show that the subset $C \subseteq \mathbf{R}^{n+1}$ given by

$$C = \{(a_0, \ldots, a_{n-1}, t) \in \mathbf{R}^{n+1} \mid t^n + a_{n-1}t^{n-1} + \cdots + a_1 t + a_0 = 0\},$$

a part of which is illustrated for $n = 2$ in Figure 2.12, is a smooth submanifold. (Hint: express C as a graph of a real-valued smooth function and extend Example 2.5.3 to cover such graphs in general.)

Exercise 2.5.6 The subset $K = \{(p, |p|) \mid p \in \mathbf{R}\} \subseteq \mathbf{R}^2$ is **not** a smooth submanifold.

Note 2.5.7 If $N \subseteq M$ is a smooth submanifold and $\dim(M) = \dim(N)$ then $N \subseteq M$ is an open subset (called an *open submanifold*). Otherwise $\dim(M) > \dim(N)$.

Example 2.5.8 Let $M_n \mathbf{R}$ be the set of $n \times n$ matrices. This is a smooth manifold since it is homeomorphic to $\mathbf{R}^{n^2}$. The subset $\mathrm{GL}_n(\mathbf{R}) \subseteq M_n \mathbf{R}$ of invertible matrices is an open submanifold (the determinant function $\det \colon M_n \mathbf{R} \to \mathbf{R}$ is continuous, so the inverse image $\mathrm{GL}_n(\mathbf{R}) = \det^{-1}(\mathbf{R} - \{0\})$ of the open set $\mathbf{R} \setminus \{0\} \subseteq \mathbf{R}$ is open).

If V is an n-dimensional vector space, let $GL(V)$ be the set of linear isomorphisms $\alpha \colon V \cong V$. By representing any linear isomorphism of $\mathbf{R}^n$ in terms of the standard basis, we may identify $GL(\mathbf{R}^n)$ and $GL_n(\mathbf{R})$.

Any linear isomorphism $f \colon V \cong W$ gives a bijection $GL(f) \colon GL(V) \cong GL(W)$ sending $\alpha \colon V \cong V$ to $f\alpha f^{-1} \colon W \cong W$. Hence, any linear isomorphism $f \colon V \cong \mathbf{R}^n$ (i.e., a choice of basis) gives a bijection $GL(f) \colon GL(V) \cong GL_n\mathbf{R}$, and hence a smooth manifold structure on $GL(V)$ (with a diffeomorphism to the open subset $GL_n\mathbf{R}$ of Euclidean n^2-space).

Prove that the smooth structure on $GL(V)$ does not depend on the choice of $f \colon V \cong \mathbf{R}^n$.

If $h \colon V \cong W$ is a linear isomorphism, prove that $GL(h) \colon GL(V) \cong GL(W)$ is a diffeomorphism respecting composition and the identity element.

Example 2.5.10 Let $M_{m \times n}\mathbf{R}$ be the set of $m \times n$ matrices (if $m = n$ we write $M_n(\mathbf{R})$ instead of $M_{n \times n}(\mathbf{R})$). This is a smooth manifold since it is homeomorphic to $\mathbf{R}^{mn}$. Let $0 \le r \le \min(m, n)$. That a matrix has rank r means that it has an $r \times r$ invertible submatrix, but no larger invertible submatrices.

The subset $M_{m \times n}^r(\mathbf{R}) \subseteq M_{m \times n}\mathbf{R}$ of matrices of rank r is a submanifold of codimension $(n - r)(m - r)$. Since some of the ideas will be valuable later on, we spell out a proof.

For the sake of simplicity, we treat the case where our matrices have an invertible $r \times r$ submatrix in the upper left-hand corner. The other cases are covered in a similar manner, taking care of indices (or by composing the chart we give below with a diffeomorphism on $M_{m \times n}\mathbf{R}$ given by multiplying with permutation matrices so that the invertible submatrix is moved to the upper left-hand corner).

So, consider the open set U of matrices

$$X = \begin{bmatrix} A & B \\ C & D \end{bmatrix}$$

with $A \in M_r(\mathbf{R})$, $B \in M_{r \times (n-r)}(\mathbf{R})$, $C \in M_{(m-r) \times r}(\mathbf{R})$ and $D \in M_{(m-r) \times (n-r)}(\mathbf{R})$ such that $\det(A) \ne 0$ (i.e., such that A is in the open subset $GL_r(\mathbf{R}) \subseteq M_r(\mathbf{R})$). The matrix X has rank exactly r if and only if the last $n - r$ columns are in the span of the first r. Writing this out, this means that X is of rank r if and only if there is an $r \times (n - r)$ matrix T such that

$$\begin{bmatrix} B \\ D \end{bmatrix} = \begin{bmatrix} A \\ C \end{bmatrix} T,$$

which is equivalent to $T = A^{-1}B$ and $D = CA^{-1}B$. Hence

$$U \cap M_{m \times n}^r(\mathbf{R}) = \left\{ \begin{bmatrix} A & B \\ C & D \end{bmatrix} \in U \,\middle|\, D - CA^{-1}B = 0 \right\}.$$

The map

$$U \to GL_r(\mathbf{R}) \times M_{r \times (n-r)}(\mathbf{R}) \times M_{(m-r) \times r}(\mathbf{R}) \times M_{(m-r) \times (n-r)}(\mathbf{R}),$$

$$\begin{bmatrix} A & B \\ C & D \end{bmatrix} \mapsto (A, B, C, D - CA^{-1}B)$$

is a diffeomorphism onto an open subset of $M_r(\mathbf{R}) \times M_{r \times (n-r)}(\mathbf{R}) \times M_{(m-r) \times r}(\mathbf{R}) \times M_{(m-r) \times (n-r)}(\mathbf{R}) \cong \mathbf{R}^{mn}$, and therefore gives a chart having the desired property that $U \cap M_{m \times n}^r(\mathbf{R})$ is the set of points such that the last $(m - r)(n - r)$ coordinates vanish.

Definition 2.5.11 A smooth map $f \colon N \to M$ is an *imbedding* if the image $f(N) \subseteq M$ is a submanifold, and the induced map $N \to f(N)$ is a diffeomorphism.

Exercise 2.5.12

Prove that
$$\mathbf{C} \to M_2(\mathbf{R}),$$
$$x + iy \mapsto \begin{bmatrix} x & -y \\ y & x \end{bmatrix}$$

defines an imbedding. More generally it defines an imbedding
$$M_n(\mathbf{C}) \to M_n(M_2(\mathbf{R})) \cong M_{2n}(\mathbf{R}).$$

Show also that this imbedding sends "conjugate transpose" to "transpose" and "multiplication" to "multiplication".

Exercise 2.5.13

Show that the map
$$f \colon \mathbf{RP}^n \to \mathbf{RP}^{n+1},$$
$$[p] = [p_0, \dots, p_n] \mapsto [p, 0] = [p_0, \dots, p_n, 0]$$

is an imbedding.

Note 2.5.14 Later we will give a very efficient way of creating smooth submanifolds, getting rid of all the troubles of finding actual charts that make the subset look like $\mathbf{R}^n$ in $\mathbf{R}^{n+k}$. We shall see that if $f \colon M \to N$ is a smooth map and $q \in N$ then more often than not the inverse image
$$f^{-1}(q) = \{p \in M \mid f(p) = q\}$$

is a submanifold of M. Examples of such submanifolds are the sphere and the space of orthogonal matrices (the inverse image of the identity matrix under the map sending a matrix A to $A^{\mathrm{T}}A$, where A^{T} is A transposed).

Example 2.5.15 This is an example where we have the opportunity to use a bit of topology. Let $f \colon M \to N$ be an imbedding, where M is a (nonempty) compact n-dimensional smooth manifold and N is a connected n-dimensional smooth manifold. Then f is a diffeomorphism. This is so because $f(M)$ is compact, and hence closed, and open since it is a codimension-zero submanifold. Hence $f(M) = N$ since N is connected. But since f is an imbedding, the map $M \to f(M) = N$ is – by definition – a diffeomorphism.

Exercise 2.5.16 (This is an important exercise. Do it: you will need the result several times.) Let $i_1 \colon N_1 \to M_1$ and $i_2 \colon N_2 \to M_2$ be smooth imbeddings and let $f \colon N_1 \to N_2$ and $g \colon M_1 \to M_2$ be continuous maps such that $i_2 f = g i_1$ (i.e., the diagram

$$
\begin{array}{ccc}
N_1 & \xrightarrow{\ f\ } & N_2 \\
i_1 \downarrow & & i_2 \downarrow \\
M_1 & \xrightarrow{\ g\ } & M_2
\end{array}
$$

commutes). Show that if g is smooth, then f is smooth.

Exercise 2.5.17 Show that the composite of two imbeddings is an imbedding.

Exercise 2.5.18 Let $0 < m \leq n$ and define the *Milnor manifold* by

$$
H(m, n) = \left\{ ([p], [q]) \in \mathbf{RP}^m \times \mathbf{RP}^n \mid \sum_{k=0}^{m} p_k q_k = 0 \right\}.
$$

Prove that $H(m, n) \subseteq \mathbf{RP}^m \times \mathbf{RP}^n$ is a smooth $(m + n - 1)$-dimensional submanifold.

Note 2.5.19 The Milnor manifolds and their complex counterparts are particularly important manifolds, because they in a certain sense give the building blocks for all manifolds (up to a certain equivalence relation called *cobordism*).

2.6 Products and Sums

Definition 2.6.1 Let $(M, \mathcal{U})$ and $(N, \mathcal{V})$ be smooth manifolds. The *(smooth) product* is the smooth manifold you get by giving the product $M \times N$ the smooth atlas given by the charts

$$
x \times y \colon U \times V \to U' \times V',
$$
$$
(p, q) \mapsto (x(p), y(q)),
$$

where $(x, U) \in \mathcal{U}$ and $(y, V) \in \mathcal{V}$.

Exercise 2.6.2 Check that this definition makes sense.

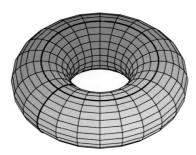

Figure 2.13. The torus is a product. The bolder curves in the illustration try to indicate the submanifolds $\{1\} \times S^1$ and $S^1 \times \{1\}$.

Note 2.6.3 Even if the atlases we start with are maximal, the charts of the form $x \times y$ do not form a maximal atlas on the product, but as always we can consider the associated maximal atlas.

Example 2.6.4 We know a product manifold already: the *torus* $S^1 \times S^1$ (Figure 2.13).

Exercise 2.6.5 Show that the projection

$$\mathrm{pr}_1 \colon M \times N \to M,$$
$$(p, q) \mapsto p$$

is a smooth map. Choose a point $p \in M$. Show that the map

$$i_p \colon N \to M \times N,$$
$$q \mapsto (p, q)$$

is an imbedding.

Exercise 2.6.6 Show that giving a smooth map $Z \to M \times N$ is the same as giving a pair of smooth maps $Z \to M$ and $Z \to N$. Hence we have a bijection

$$\mathcal{C}^\infty(Z, M \times N) \cong \mathcal{C}^\infty(Z, M) \times \mathcal{C}^\infty(Z, N).$$

Exercise 2.6.7 Show that the infinite cylinder $\mathbf{R}^1 \times S^1$ is diffeomorphic to $\mathbf{R}^2 \setminus \{0\}$. See Figure 2.14. More generally: $\mathbf{R}^1 \times S^n$ is diffeomorphic to $\mathbf{R}^{n+1} \setminus \{0\}$.

Exercise 2.6.8 Let $f \colon M \to M'$ and $g \colon N \to N'$ be imbeddings. Then

$$f \times g \colon M \times N \to M' \times N'$$

is an imbedding.

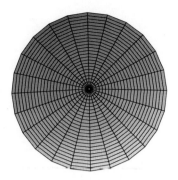

Figure 2.14. Looking down into the infinite cylinder.

Exercise 2.6.9

Show that there exists an imbedding $S^{n_1} \times \cdots \times S^{n_k} \to \mathbf{R}^{1 + \sum_{i=1}^{k} n_i}$.

Exercise 2.6.10

Why is the multiplication of matrices

$$\mathrm{GL}_n(\mathbf{R}) \times \mathrm{GL}_n(\mathbf{R}) \to \mathrm{GL}_n(\mathbf{R}), \qquad (A, B) \mapsto A \cdot B$$

a smooth map? This, together with the existence of inverses, makes $\mathrm{GL}_n(\mathbf{R})$ a "*Lie group*".

For the record: a *Lie group* is a smooth manifold M with a smooth "multiplication" $M \times M \to M$ that is associative, and it has a neutral element and all inverses (in $\mathrm{GL}_n(\mathbf{R})$ the neutral element is the identity matrix).

Exercise 2.6.11

Why is the multiplication

$$S^1 \times S^1 \to S^1, \qquad (e^{i\theta}, e^{i\tau}) \mapsto e^{i\theta} \cdot e^{i\tau} = e^{i(\theta + \tau)}$$

a smooth map? This is our second example of a Lie group.

Definition 2.6.12 Let $(M, \mathcal{U})$ and $(N, \mathcal{V})$ be smooth manifolds. The *(smooth) disjoint union* (or *sum*) is the smooth manifold you get by giving the disjoint union $M \coprod N$ the smooth structure given by $\mathcal{U} \cup \mathcal{V}$. See Figure 2.15.

Exercise 2.6.13

Check that this definition makes sense.

Note 2.6.14 As for the product, the atlas we give the sum is not maximal (a chart may have disconnected source and target). There is nothing wrong a priori with taking the disjoint union of an m-dimensional manifold with an n-dimensional manifold. The result will of course be neither m- nor n-dimensional. Such examples will not be important to us, and you will find that in arguments we may talk about a smooth manifold, and without hesitation later on start talking about its dimension. This is justified since we can consider one component at a time, and each component will have a well-defined dimension.

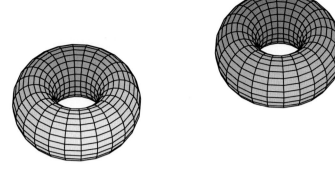

Figure 2.15. The disjoint union of two tori (imbedded in $\mathbf{R}^3$).

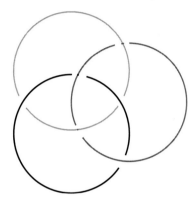

Figure 2.16.

Note 2.6.15 Manifolds with boundary, as defined in Note 2.3.17, do not behave nicely under products, as can be seen already from the simplest example $[0, 1] \times [0, 1]$. There are ways of "rounding the corners" that have been used in the literature to deal with this problem, but it soon becomes rather technical.

There is no similar problem with the disjoint union.

Example 2.6.16 The Borromean rings (Figure 2.16) give an interesting example showing that the imbedding in Euclidean space is irrelevant to the manifold: the Borromean rings amount to the disjoint union of three circles $S^1 \coprod S^1 \coprod S^1$. Don't get confused: it is the imbedding in $\mathbf{R}^3$ that makes your mind spin: the manifold itself is just three copies of the circle! Moral: an imbedded manifold is something more than just a manifold that *can be* imbedded.

Exercise 2.6.17 Prove that the inclusion

$$\mathrm{inc}_1 \colon M \subset M \coprod N$$

is an imbedding.

Exercise 2.6.18 Show that giving a smooth map $M \coprod N \to Z$ is the same as giving a pair of smooth maps $M \to Z$ and $N \to Z$. Hence we have a bijection

$$C^\infty(M \coprod N, Z) \cong C^\infty(M, Z) \times C^\infty(N, Z).$$

Exercise 2.6.19 Find all the errors in the hints in Appendix B for the exercises from Chapter 2.

3 The Tangent Space

In this chapter we will study *linearizations*. You have seen this many times before as *tangent lines* and *tangent planes* (for curves and surfaces in Euclidean space), and the main difficulty you will encounter is that the linearizations must be defined intrinsically – i.e., in terms of the manifold at hand – and not with reference to some big ambient space. We will shortly (in Predefinition 3.0.5) give a simple and perfectly fine technical definition of the tangent space, but for future convenience we will use the concept of *germs* in our final definition. This concept makes notation and bookkeeping easy and is good for all things local (in the end it will turn out that due to the existence of so-called smooth bump functions (see Section 3.2) we could have stayed global in our definitions).

An important feature of the tangent space is that it is a vector space, and a smooth map of manifolds gives a linear map of vector spaces. Eventually, the chain rule expresses the fact that the tangent space is a "natural" construction (which actually is a very precise statement that will reappear several times in different contexts. It is the hope of the author that the reader, through the many examples, in the end will appreciate the importance of being natural – as well as earnest).

Beside the tangent space, we will also briefly discuss its sibling, the *cotangent space*, which is concerned with linearizing the space of real-valued functions, and which is the relevant linearization for many applications.

Another interpretation of the tangent space is as the space of *derivations*, and we will discuss these briefly since they figure prominently in many expositions. They are more abstract and less geometric than the path we have chosen – as a matter of fact, in our presentation derivations are viewed as a "double dualization" of the tangent space.

3.0.1 The Idea of the Tangent Space of a Submanifold of Euclidean Space

Given a submanifold M of Euclidean space $\mathbf{R}^n$, it is fairly obvious what we should mean by the "tangent space" of M at a point $p \in M$.

In purely physical terms, the tangent space should be the following subspace of $\mathbf{R}^n$. If a particle moves on some curve in M and at p suddenly "loses its grip on M" (Figure 3.1) it will continue out in the ambient space along a straight line (its "tangent"). This straight line is determined by its velocity vector at the point

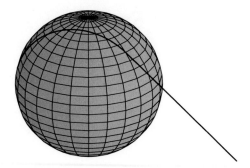

Figure 3.1. A particle loses its grip on M and flies out on a tangent

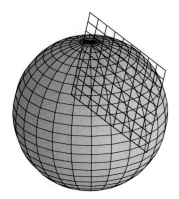

Figure 3.2. A part of the space of all tangents

where it flies out into space. The tangent space should be the linear subspace of $\mathbf{R}^n$ containing all these vectors. See Figure 3.2.

When talking about manifolds it is important to remember that there **is no** ambient space to fly out into, but we still may talk about a tangent space.

3.0.2 Partial Derivatives

The tangent space is all about the linearization in Euclidean space. To fix notation we repeat some multivariable calculus.

Definition 3.0.1 Let $f : U \to \mathbf{R}$ be a function where U is an open subset of $\mathbf{R}^n$ containing $p = (p_1, \ldots p_n)$. The *ith partial derivative of f at p* is the number (if it exists)

$$D_i f(p) = D_i|_p\, f = \lim_{h \to 0} \frac{1}{h}(f(p + h e_i) - f(p)),$$

where e_i is the ith unit vector $e_i = (0, \ldots, 0, 1, 0, \ldots, 0)$ (with a 1 in the ith coordinate). We collect the partial derivatives in a $1 \times n$ matrix

$$Df(p) = D|_p f = (D_1 f(p), \ldots, D_n f(p)).$$

Definition 3.0.2 If $f = (f_1, \ldots, f_m): U \to \mathbf{R}^m$ is a function where U is an open subset of $\mathbf{R}^n$ containing $p = (p_1, \ldots p_n)$, then the *Jacobian matrix* is the $m \times n$ matrix

$$Df(p) = D|_p(f) = \begin{bmatrix} Df_1(p) \\ \vdots \\ Df_m(p) \end{bmatrix}.$$

In particular, if $g = (g_1, \ldots g_m): (a, b) \to \mathbf{R}^m$ the Jacobian is an $m \times 1$ matrix, or element in $\mathbf{R}^m$, which we write as

$$g'(c) = Dg(c) = \begin{bmatrix} g_1'(c) \\ \vdots \\ g_m'(c) \end{bmatrix} \in \mathbf{R}^m.$$

For convenience, we cite the "flat" (i.e., in Euclidean space) chain rule. For a proof, see, e.g., Section 2-9 of [19], or any decent book on multivariable calculus.

Lemma 3.0.3 *(The Flat Chain Rule) Let $g: (a, b) \to U$ and $f: U \to \mathbf{R}$ be smooth functions where U is an open subset of $\mathbf{R}^n$ and $c \in (a, b)$. Then*

$$(fg)'(c) = D(f)(g(c)) \cdot g'(c)$$
$$= \sum_{j=1}^{n} D_j f(g(c)) \cdot g_j'(c).$$

Note 3.0.4 When considered as a vector space, we insist that the elements in $\mathbf{R}^n$ are standing vectors (so that linear maps can be represented by multiplication by matrices from the left); when considered as a manifold the distinction between lying and standing vectors is not important, and we use either convention as may be typographically convenient.

It is a standard fact from multivariable calculus (see, e.g., Section 2-8 of [19]) that if $f: U \to \mathbf{R}^m$ is continuously differentiable at p (all the partial derivatives exist and are continuous at p), where U is an open subset of $\mathbf{R}^n$, then the Jacobian is the matrix associated (in the standard bases) with the unique linear transformation $L: \mathbf{R}^n \to \mathbf{R}^m$ such that

$$\lim_{h \to 0} \frac{1}{|h|} (f(p + h) - f(p) - L(h)) = 0.$$

Predefinition 3.0.5 (of the Tangent Space) *Let M be a smooth manifold, and let $p \in M$. Consider the set of all curves $\gamma: \mathbf{R} \to M$ with $\gamma(0) = p$. On this set we define the following equivalence relation: given two curves $\gamma: \mathbf{R} \to M$ and*

$\gamma_1 \colon \mathbf{R} \to M$ with $\gamma(0) = \gamma_1(0) = p$, we say that γ and γ_1 are equivalent if for all charts $x \colon U \to U'$ with $p \in U$ we have an equality of vectors

$$(x\gamma)'(0) = (x\gamma_1)'(0).$$

Then the tangent space of M at p is the set of all equivalence classes.

There is nothing wrong with this definition, in the sense that it is naturally iso-morphic to the one we are going to give in a short while (see Definition 3.3.1). However, in order to work efficiently with our tangent space, it is fruitful to intro-duce some language. It is really not necessary for our curves to be defined on all of $\mathbf{R}$, but on the other hand it is not important to know the domain of definition as long as it contains a neighborhood around the origin.

3.1 Germs

Whatever one's point of view on tangent vectors is, it is a local concept. The tangent of a curve passing through a given point p is only dependent upon the behavior of the curve close to the point. Hence it makes sense to divide out by the equivalence relation which says that all curves that are equal on some neighborhood of the point are equivalent. This is the concept of *germs*.

Definition 3.1.1 Let M and N be smooth manifolds, and let $p \in M$. On the set

$$\{f \mid f \colon U_f \to N \text{ is smooth, and } U_f \text{ an open neighborhood of } p\}$$

we define an equivalence relation where f is equivalent to g, written $f \sim g$, if there is an open neighborhood $V_{fg} \subseteq U_f \cap U_g$ of p such that

$$f(q) = g(q), \text{ for all } q \in V_{fg}.$$

Such an equivalence class is called a *germ*, and we write

$$\bar{f} \colon (M, p) \to (N, f(p))$$

for the germ associated with $f \colon U_f \to N$. We also say that f *represents* $\bar{f}$.

Definition 3.1.2 Let M be a smooth manifold and p a point in M. A *function germ* at p is a germ $\bar{\phi} \colon (M, p) \to (\mathbf{R}, \phi(p))$. Let

$$\mathcal{O}_{M,p} = \mathcal{O}_p$$

be the set of function germs at p.

Example 3.1.3 In $\mathcal{O}_{\mathbf{R}^n, 0}$ there are some very special function germs, namely those associated with the *standard coordinate functions* pr_i sending $p = (p_1, \ldots, p_n)$ to $\mathrm{pr}_i(p) = p_i$ for $i = 1, \ldots, n$.

Note 3.1.4 Germs are quite natural things. Most of the properties we need about germs are "obvious" if you do not think too hard about them, so it is a good idea to skip the rest of the section which spells out these details before you know what they are good for. Come back later if you need anything precise.

Exercise 3.1.5 Show that the relation $\sim$ actually is an equivalence relation as claimed in Definition 3.1.1.

Let

$$\bar{f}\colon (M, p) \to (N, f(p))$$

and

$$\bar{g}\colon (N, f(p)) \to (L, g(f(p)))$$

be two germs represented by the functions $f\colon U_f \to N$ and $g\colon U_g \to L$. Then we define the *composite*

$$\bar{g}\,\bar{f}\colon (M, p) \to (L, g(f(p)))$$

as the germ associated with the composite

$$f^{-1}(U_g) \xrightarrow{\;f|_{f^{-1}(U_g)}\;} U_g \xrightarrow{\;g\;} L$$

(which makes sense since $f^{-1}(U_g) \subseteq M$ is an open set containing p). See Figure 3.3.

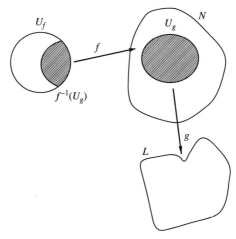

Figure 3.3. The composite of two germs: just remember to restrict the domain of the representatives.

Exercise 3.1.6

Show that the composition $\bar{g}\bar{f}$ of germs is well defined in the sense that it does not depend on the chosen representatives g and f. Also, show "associativity": $\bar{h}(\bar{g}\bar{f}) = (\bar{h}\bar{g})\bar{f}$; and that, if $\bar{h}$ and $\bar{f}$ are represented by identity functions, then $\bar{h}\bar{g} = \bar{g} = \bar{g}\bar{f}$.

We occasionally write $\overline{gf}$ instead of $\bar{g}\bar{f}$ for the composite, even though pedants will point out that we have to adjust the domains before composing representatives.

Also, we will be cavalier about the range of germs, in the sense that if $q \in V \subseteq N$ we will sometimes not distinguish notationally between a germ $(M, p) \to (V, q)$ and the germ $(M, p) \to (N, q)$ given by composition with the inclusion.

A germ $\bar{f}\colon (M, p) \to (N, q)$ is *invertible* if (and only if) there is a germ $\bar{g}\colon (N, q) \to (M, p)$ such that the composites $\bar{f}\bar{g}$ and $\bar{g}\bar{f}$ are represented by identity maps.

Lemma 3.1.7 *A germ $\bar{f}\colon (M, p) \to (N, q)$ represented by $f\colon U_f \to N$ is invertible if and only if there is a diffeomorphism $\phi\colon U \to V$ with $U \subseteq U_f$ an open neighborhood of p and V an open neighborhood of q such that $f(t) = \phi(t)$ for all $t \in U$.*

Exercise 3.1.8

Prove Lemma 3.1.7. A full proof is provided in Appendix B, but . . .

Note 3.1.9 The set $\mathcal{O}_{M,p}$ of function germs forms a vector space by pointwise addition and multiplication by real numbers:

$$\bar{\phi} + \bar{\psi} = \overline{\phi + \psi}, \qquad \text{where } (\phi + \psi)(q) = \phi(q) + \psi(q) \text{ for } q \in U_\phi \cap U_\psi,$$
$$k \cdot \bar{\phi} = \overline{k \cdot \phi}, \qquad \text{where } \quad (k \cdot \phi)(q) = k \cdot \phi(q) \qquad \text{for } q \in U_\phi,$$
$$\bar{0}, \qquad \text{where} \qquad 0(q) = 0 \qquad \text{for } q \in M.$$

It furthermore has the pointwise multiplication, making it what is called a "commutative **R**-algebra":

$$\bar{\phi} \cdot \bar{\psi} = \overline{\phi \cdot \psi}, \qquad \text{where } (\phi \cdot \psi)(q) = \phi(q) \cdot \psi(q) \text{ for } q \in U_\phi \cap U_\psi,$$
$$\bar{1}, \qquad \text{where} \qquad 1(q) = 1 \qquad \text{for } q \in M.$$

That these structures obey the usual rules follows by the same rules on **R**.

Since we both multiply and compose germs, we should perhaps be careful in distinguishing the two operations by remembering to write $\circ$ whenever we compose, and $\cdot$ when we multiply. We will be sloppy about this, and the $\circ$ will mostly be invisible. We try to remember to write the $\cdot$, though.

Definition 3.1.10 A germ $\bar{f}\colon (M, p) \to (N, f(p))$ defines a function

$$f^*\colon \mathcal{O}_{f(p)} \to \mathcal{O}_p$$

by sending a function germ $\bar{\phi} \colon (N, f(p)) \to (\mathbf{R}, \phi f(p))$ to

$$\overline{\phi f} \colon (M, p) \to (\mathbf{R}, \phi f(p))$$

("precomposition").

Note that f^* preserves addition and multiplication.

Lemma 3.1.11 *If* $\bar{f} \colon (M, p) \to (N, f(p))$ *and* $\bar{g} \colon (N, f(p)) \to (L, g(f(p)))$ *then*

$$f^* g^* = (gf)^* \colon \mathcal{O}_{L, g(f(p))} \to \mathcal{O}_{M, p}.$$

Exercise 3.1.12 Prove Lemma 3.1.11. A full proof is provided in Appendix B, but ...

The superscript $*$ may help you remember that this construction reverses the order, since it may remind you of transposition of matrices.

Since manifolds are locally Euclidean spaces, it is hardly surprising that, on the level of function germs, there is little difference between $(\mathbf{R}^n, 0)$ and (M, p).

Lemma 3.1.13 *There are isomorphisms* $\mathcal{O}_{M, p} \cong \mathcal{O}_{\mathbf{R}^n, 0}$ *preserving all algebraic structure.*

Proof. Pick a chart $x \colon U \to U'$ with $p \in U$ and $x(p) = 0$ (if $x(p) \neq 0$, just translate the chart). Then

$$x^* \colon \mathcal{O}_{\mathbf{R}^n, 0} \to \mathcal{O}_{M, p}$$

is invertible with inverse $(x^{-1})^*$ (note that $\overline{\mathrm{id}_U} = \overline{\mathrm{id}_M}$ since they agree on an open subset (namely U) containing p). $\blacksquare$

Note 3.1.14 So is this the end of the subject? Could we just as well study $\mathbf{R}^n$? No! these isomorphisms depend on **a choice of charts**. This is OK if you just look at one point at a time, but as soon as things get a bit messier, this is every bit as bad as choosing particular coordinates in vector spaces.

3.2 Smooth Bump Functions

Germs allow us to talk easily about local phenomena. There is another way of focusing our attention on neighborhoods of a point p in a smooth manifold M, namely by using bump functions. Their importance lies in the fact that they can represent all local features near p while ignoring everything "far away". The existence of smooth bump functions is a true luxury about smooth manifolds, which makes the smooth case much more flexible than the analytic case. We will return to this topic when we define partitions of unity.

Definition 3.2.1 Let X be a space and p a point in X. A *bump function* around p is a map $\phi \colon X \to \mathbf{R}$, which takes values in the closed interval $[0, 1]$ only, which takes the constant value 1 in (the closure of) a neighborhood of p, and takes the constant value 0 outside some bigger neighborhood.

We will be interested only in *smooth* bump functions.

Definition 3.2.2 Let X be a space. The *support* of a function $f \colon X \to \mathbf{R}$ is the closure of the subset of X with nonzero values, i.e.,

$$\operatorname{supp}(f) = \overline{\{x \in X \mid f(x) \neq 0\}}.$$

Lemma 3.2.3 *Given $r, \epsilon > 0$, there is a smooth bump function*

$$\gamma_{r,\epsilon} \colon \mathbf{R}^n \to \mathbf{R}$$

with $\gamma_{r,\epsilon}(t) = 1$ for $|t| \leq r$ and $\gamma_{r,\epsilon}(t) = 0$ for $|t| \geq r + \epsilon$. More generally, if M is a manifold and $p \in M$, then there exist smooth bump functions around p. See Figure 3.4.

Proof. Let $\beta_\epsilon \colon \mathbf{R} \to \mathbf{R}$ be any smooth function with non-negative values and support $[0, \epsilon]$ (for instance, you may use $\beta_\epsilon(t) = \lambda(t) \cdot \lambda(\epsilon - t)$, where λ is the function of Exercise 2.2.14).

Since β_ϵ is smooth, it is integrable with $\int_0^\epsilon \beta_\epsilon(x) dx > 0$, and we may define the smooth step function $\alpha_\epsilon \colon \mathbf{R} \to \mathbf{R}$ which ascends from zero to one smoothly between zero and ϵ by means of

$$\alpha_\epsilon(t) = \frac{\int_0^t \beta_\epsilon(x) dx}{\int_0^\epsilon \beta_\epsilon(x) dx}.$$

See Figure 3.5. Finally, $\gamma_{(r,\epsilon)} \colon \mathbf{R}^n \to \mathbf{R}$ is given by

$$\gamma_{(r,\epsilon)}(x) = 1 - \alpha_\epsilon(|x| - r).$$

As to the more general case, choose a chart (x, U) for the smooth manifold M with $p \in U$. By translating, we may assume that $x(p) = 0$. Since $x(U) \subseteq \mathbf{R}^n$

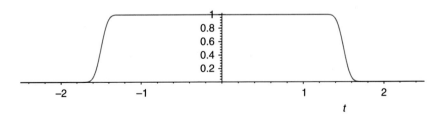

Figure 3.4.

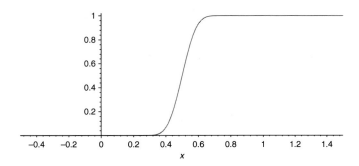

Figure 3.5.

is open, there are $r, \epsilon > 0$ such that the open ball of radius $r + 2\epsilon$ is contained in $x(U)$. The function given by sending $q \in M$ to $\gamma_{(r,\epsilon)}x(q)$ if $q \in U$ and to 0 if $q \neq U$ is a smooth bump function around p. ∎

Example 3.2.4 Smooth bump functions are very handy, for instance if you want to join curves in a smooth fashion (e.g. if you want to design smooth highways!). They also allow you to drive smoothly on a road with corners: the curve $\gamma : \mathbf{R} \to \mathbf{R}^2$ given by $\gamma(t) = (te^{-1/t^2}, |te^{-1/t^2}|)$ is smooth, although its image is not a smooth submanifold.

Exercise 3.2.5 Given $\epsilon > 0$, prove that there is a diffeomorphism $f : (-\epsilon, \epsilon) \to \mathbf{R}$ such that $f(t) = t$ for $|t|$ small. Conclude that any germ $\bar{\gamma} : (\mathbf{R}, 0) \to (M, p)$ is represented by a "globally defined" curve $\gamma : \mathbf{R} \to M$.

Exercise 3.2.6 Show that any function germ $\bar{\phi} : (M, p) \to (\mathbf{R}, \phi(p))$ has a smooth representative $\phi : M \to \mathbf{R}$.

Exercise 3.2.7 Let M and N be smooth manifolds and $f : M \to N$ a continuous map. Show that f is smooth if for all smooth $\phi : N \to \mathbf{R}$ the composite $\phi f : M \to \mathbf{R}$ is smooth.

3.3 The Tangent Space

Note that if $\bar{\gamma} : (\mathbf{R}, 0) \to (\mathbf{R}^n, \gamma(0))$ is some germ into Euclidean space, the derivative at zero does not depend on a choice of representative (i.e., if γ and γ_1 are two representatives for $\bar{\gamma}$, then $\gamma'(0) = \gamma_1'(0)$), and we write $\gamma'(0)$ without ambiguity.

Definition 3.3.1 Let $(M, \mathcal{A})$ be a smooth n-dimensional manifold. Let $p \in M$ and let

$$W_p = \{\text{germs } \bar{\gamma} : (\mathbf{R}, 0) \to (M, p)\}.$$

Two germs $\bar{\gamma}, \bar{\gamma}_1 \in W_p$ are said to be equivalent, written $\bar{\gamma} \approx \bar{\gamma}_1$, if for all function germs $\bar{\phi} : (M, p) \to (\mathbf{R}, \phi(p))$ we have that $(\phi\gamma)'(0) = (\phi\gamma_1)'(0)$. We define the *tangent space* of M at p to be the set of equivalence classes

$$T_p M = W_p / \approx .$$

We write $[\bar{\gamma}]$ (or simply $[\gamma]$) for the $\approx$-equivalence class of $\bar{\gamma}$. This definition is essentially the same as the one we gave in Predefinition 3.0.5 (see Lemma 3.3.9 below). So, for the definition of the tangent space, it is not necessary to involve the definition of germs, but it is convenient when working with the definition since we are freed from specifying domains of definition all the time.

Exercise 3.3.2 Show that the equivalence relation on W_p in Definition 3.3.1 could equally well be described as follows. Two germs $\bar{\gamma}, \bar{\gamma}_1 \in W_p$ are said to be equivalent, if for all charts $(x, U) \in \mathcal{A}$ with $p \in U$ we have that $(x\gamma)'(0) = (x\gamma_1)'(0)$.

Exercise 3.3.3 Show that, for two germs $\bar{\gamma}, \bar{\gamma}_1 : (\mathbf{R}, 0) \to (M, p)$ to define the same tangent vector, it suffices that $(x\gamma)'(0) = (x\gamma_1)'(0)$ for *some* chart (x, U).

However – as is frequently the case – it is not the *objects*, but the *maps* comparing them, that turn out to be most important. Hence we need to address how the tangent space construction is to act on smooth maps and germs.

Definition 3.3.4 Let $\bar{f} : (M, p) \to (N, f(p))$ be a germ. Then we define

$$T_p f : T_p M \to T_{f(p)} N$$

by

$$T_p f([\gamma]) = [f\gamma].$$

Exercise 3.3.5 This is well defined.

Does anybody recognize the next lemma? It is the chain rule!

Lemma 3.3.6 *If* $\bar{f} : (M, p) \to (N, f(p))$ *and* $\bar{g} : (N, f(p)) \to (L, g(f(p)))$ *are germs, then*

$$T_{f(p)} g \, T_p f = T_p(gf).$$

Proof.   Let $\bar{\gamma} \colon (\mathbf{R}, 0) \to (M, p)$, then

$$T_{f(p)}g(T_p f([\gamma])) = T_{f(p)}g([f\gamma]) = [gf\gamma] = T_p(gf)([\gamma]).\qquad\blacksquare$$

That's the ultimate proof of the chain rule! The ultimate way to remember it is as follows: the two ways around the triangle

$$
\begin{array}{ccc}
T_p M & \xrightarrow{\ T_p f\ } & T_{f(p)} N \\[2pt]
& \searrow\rlap{\scriptstyle T_p(gf)} & \big\downarrow{\scriptstyle T_{f(p)}g} \\[6pt]
& & T_{gf(p)} L
\end{array}
$$

are the same ("the diagram commutes").

Note 3.3.7   For the categorists: the tangent space is an assignment from pointed manifolds to vector spaces, and the chain rule states that it is a "functor".

Exercise 3.3.8   Show that, if the germ $\bar{f} \colon (M, p) \to (N, f(p))$ is invertible (i.e., there is a germ $\bar{g} \colon (N, f(p)) \to (M, p)$ such that $\bar{g}\bar{f}$ is the identity germ on (M, p) and $\bar{f}\bar{g}$ is the identity germ on $(N, f(p))$), then $T_p f$ is a bijection with inverse $T_{f(p)}g$. In particular, the tangent space construction sends diffeomorphisms to bijections.

3.3.1 The Vector Space Structure

The "flat chain rule" in Lemma 3.0.3 from multivariable calculus will be used to show that the tangent spaces are vector spaces and that $T_p f$ is a linear map, but, if we were content with working with sets only, the one-line proof of the chain rule in Lemma 3.3.6 would be all we'd ever need.

Summing up the Exercises 3.3.2 and 3.3.3, the predefinition of the tangent space given in Predefinition 3.0.5 agrees with the official definition (we allow ourselves to make the conclusion of exercises official when full solutions are provided).

Proposition 3.3.9   *The tangent space at a point p is the set of all (germs of) curves sending 0 to p, modulo the identification of all curves having equal derivatives at 0 in some chart.*

Proof.   This is the contents of the Exercises 3.3.2 and 3.3.3, and, since by Exercise 3.2.5 all germs of curves have representatives defined on all of $\mathbf{R}$, the parenthesis could really be removed. $\blacksquare$

In particular if $M = \mathbf{R}^n$, then two curves $\gamma_1, \gamma_2 \colon (\mathbf{R}, 0) \to (\mathbf{R}^n, p)$ define the same tangent vector if and only if the derivatives are equal:

$$\gamma_1'(0) = \gamma_2'(0)$$

(using the identity chart). Hence, a tangent vector in $\mathbf{R}^n$ is uniquely determined by (p and) its derivative at 0, and so $T_p \mathbf{R}^n$ may be identified with $\mathbf{R}^n$. See Figure 3.6.

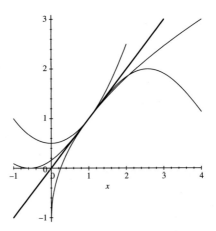

Figure 3.6. Many curves give rise to the same tangent.

Lemma 3.3.10   *A germ $\bar{\gamma}: (\mathbf{R}, 0) \to (\mathbf{R}^n, p)$ is $\approx$-equivalent to the germ represented by*

$$t \mapsto p + \gamma'(0)t.$$

That is, all elements in $T_p\mathbf{R}^n$ are represented by linear curves, giving a bijection

$$T_p\mathbf{R}^n \cong \mathbf{R}^n, \qquad [\gamma] \mapsto \gamma'(0).$$

More generally, if M is an n-dimensional smooth manifold, p a point in M and (x, U) a chart with $p \in U$, then the map

$$A_x: T_pM \to \mathbf{R}^n, \qquad A_x([\gamma]) = (x\gamma)'(0)$$

is a bijection with inverse $A_x^{-1}(v) = [B_x^v]$, where $B_x^v(t) = x^{-1}(x(p) + tv)$.

Proof.   It suffices to check that the purported formula for the inverse actually works. We check both composites, using that $xB_x^v(t) = x(p) + tv$, and so $(xB_x^v)'(0) = v$: $A_x^{-1}A_x([\gamma]) = [B_x^{(x\gamma)'(0)}] = [\gamma]$ and $A_xA_x^{-1}(v) = (xB_x^v)'(0) = v$. ■

Proposition 3.3.11   *Let $\bar{f}: (M, p) \to (N, f(p))$ be a germ, (x, U) a chart in M with $p \in U$ and (y, V) a chart in N with $f(p) \in V$. Then the diagram*

$$
\begin{array}{ccc}
T_pM & \xrightarrow{\;\;T_pf\;\;} & T_{f(p)}N \\[2pt]
\cong \downarrow A_x & & \cong \downarrow A_y \\[2pt]
\mathbf{R}^m & \xrightarrow{\;D(yfx^{-1})(x(p))\cdot\;} & \mathbf{R}^n
\end{array}
$$

commutes, where the bottom horizontal map is the linear map given by multiplication with the Jacobian matrix $D(yfx^{-1})(x(p))$ (c.f. Definition 3.0.2).

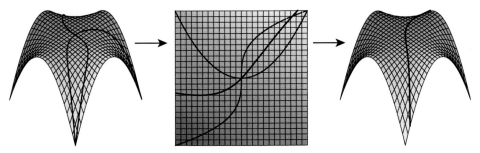

Figure 3.7. Two curves on M are sent by a chart x to $\mathbf{R}^n$, where they are added, and the sum is sent back to M with x^{-1}.

Proof. To show that the diagram commutes, start with a $[\gamma] \in T_pM$. Going down and right we get $D(yfx^{-1})(x(p)) \cdot (x\gamma)'(0)$ and going right and down we get $(yf\gamma)'(0)$. That these two expressions agree is the chain rule: $(yf\gamma)'(0) = (yfx^{-1}x\gamma)'(0) = D(yfx^{-1})(x(p)) \cdot (x\gamma)'(0)$. ∎

Proposition 3.3.11 is extremely useful, not only because it eventually will prove that the tangent map T_pf in a natural way is linear (we define things so that A_x and A_y turn out to be linear isomorphisms, giving that T_pf is displayed as the composite of three linear maps), but also because it gives us a concrete way of *calculating* the tangent map. Many questions can be traced back to a question of whether T_pf is onto ("p is a regular point"), and we see that Proposition 3.3.11 translates this to the question of whether the Jacobi matrix $D(yfx^{-1})(x(p))$ has rank equal to the dimension of N.

Note 3.3.12 The tangent space is a vector space, and like always we fetch the structure locally by means of charts. Visually it goes like the sequence shown in Figure 3.7.

Explicitly, if $[\gamma_1], [\gamma_2] \in T_pM$ and $a, b \in \mathbf{R}$ we define

$$a[\gamma_1] + b[\gamma_2] = A_x^{-1}(aA_x[\gamma_1] + bA_x[\gamma_2]).$$

This is all well and fine, but would have been quite worthless if the vector space structure depended on a choice of chart. Of course, it does not. To see this we use the following observation.

Exercise 3.3.13 Let X be a set, V, W vector spaces, $\alpha: X \to V$ a bijection and $\beta: V \to W$ a linear isomorphism. Then show that the vector space structures induced on X by α and $\beta\alpha$ are equal. Note that α becomes a linear isomorphism with this structure.

Lemma 3.3.14 *The vector space structure on T_pM in Note 3.3.12 is independent of the choice of chart. Furthermore, if $\bar{f}: (M, p) \to (N, q)$ is a germ, then $T_pf: T_pM \to T_qN$ is linear.*

Proof.   Consider the diagram in Proposition 3.3.11 when f is the identity:

$$
\begin{array}{ccc}
T_pM & =\!=\!=\!= & T_pM \\
\cong \downarrow A_x & & \cong \downarrow A_y \\
\mathbf{R}^m & \xrightarrow{D(yx^{-1})(x(p))\cdot} & \mathbf{R}^m
\end{array}
$$

with two potentially different charts (x, U) and (y, V). Since $\bar{y}\bar{x}^{-1}$ is invertible, the Jacobi matrix $D(yx^{-1})(x(p))$ is invertible and so we see that the bijection A_y is equal to the composite of A_x with a linear isomorphism. Hence, by the observation in Exercise 3.3.13 (with $\alpha = A_x$ and β multiplication by the Jacobi matrix), the vector space structure induced on T_pM by A_x is equal to the vector space structure induced by A_y.

That the tangent map is linear now follows immediately from the diagram in Proposition 3.3.11 for a general $\bar{f}$. ■

Example 3.3.15   Consider the map det: $M_2(\mathbf{R}) \to \mathbf{R}$ sending the matrix

$$
A = \begin{bmatrix} a_{11} & a_{12} \\ a_{21} & a_{22} \end{bmatrix}
$$

to its determinant $\det(A) = a_{11}a_{22} - a_{12}a_{21}$. Using the chart $x: M_2(\mathbf{R}) \to \mathbf{R}^4$ with

$$
x(A) = \begin{bmatrix} a_{11} \\ a_{12} \\ a_{21} \\ a_{22} \end{bmatrix}
$$

(and the identity chart on $\mathbf{R}$) we have that the Jacobi matrix is the 1×4 matrix

$$
D(\det x^{-1})(x(A)) = [a_{22}, -a_{21}, -a_{12}, a_{11}]
$$

(check this!). Thus we see that the rank of $D(\det x^{-1})(x(A))$ is 0 if $A = 0$ and 1 if $A \neq 0$. Hence $T_A \det: T_A M_2(\mathbf{R}) \to T_{\det A}\mathbf{R}$ is onto if and only if $A \neq 0$ (and $T_0 \det = 0$).

Exercise 3.3.16   Consider the determinant map det: $M_n(\mathbf{R}) \to \mathbf{R}$ for $n > 1$. Show that $T_A \det$ is onto if the rank rk A of the $n \times n$ matrix A is greater than $n - 2$ and $T_A \det = 0$ if rk $A < n - 1$.

Exercise 3.3.17   Let $L: \mathbf{R}^n \to \mathbf{R}^m$ be a linear transformation. Show that $DL(p)$ is the matrix associated with L in the standard basis (and hence independent of the point p).

3.4 The Cotangent Space

Although the tangent space has a clear physical interpretation as the space of all possible velocities at a certain point of a manifold, it turns out that for many applications – including mechanics – the *cotangent space* is even more fundamental.

As opposed to the tangent space, which is defined in terms of maps **from** the real line to the manifold, the cotangent space is defined in turn of maps **to** the real line. We are really having a glimpse of a standard mathematical technique: if you want to understand an object, a good way is to understand the maps to or from something you think you understand (in this case the real line). The real line is the "yardstick" for spaces.

Recall from Definition 3.1.2 that $\mathcal{O}_{M,p}$ denotes the algebra of function germs $\bar\phi\colon (M, p) \to (\mathbf{R}, \phi(p))$. If W is a subspace of a vector space V, then the quotient space V/W is the vector space you get from V by dividing out by the equivalence relation $v \sim v + w$ for $v \in V$ and $w \in W$. The vector space structure on V/W is defined by demanding that the map $V \to V/W$ sending a vector to its equivalence class is linear.

Definition 3.4.1   Let M be a smooth manifold, and let $p \in M$. Let

$$J = J_p = J_{M,p} \subseteq \mathcal{O}_{M,p}$$

be the vector space of all smooth function germs $\bar\phi\colon (M, p) \to (\mathbf{R}, 0)$ (i.e., such that $\phi(p) = 0$), and let $J^2 = J_p^2 = J_{M,p}^2$ be the sub-vector space spanned by all products $\bar\phi \cdot \bar\psi$, where $\bar\phi$ and $\bar\psi$ are in J. The *cotangent space, T_p^*M, of M at p* is the quotient space

$$T_p^*M = J_p/J_p^2.$$

The elements of T_p^*M are referred to as *cotangent vectors*.

Let $\bar f\colon (M, p) \to (N, f(p))$ be a smooth germ. Then

$$T^*f = T_p^*f\colon T_{f(p)}^*N \to T_p^*M$$

is the linear transformation given by sending sending the cotangent vector represented by the function germ $\bar\psi\colon (N, f(p)) \to (\mathbf{R}, 0)$ to the cotangent vector represented by $\bar\psi \bar f\colon (M, p) \to (N, f(p)) \to (\mathbf{R}, 0)$.

Lemma 3.4.2   *If $\bar f\colon (M, g(p)) \to (N, fg(p))$ and $\bar g\colon (L, p) \to (M, g(p))$ are smooth germs, then $T^*(fg) = T^*g T^*f$, i.e.,*

$$
\begin{array}{ccc}
T_{fg(p)}^*N & \xrightarrow{\;T^*f\;} & T_{g(p)}^*M \\
& \searrow{\scriptstyle T^*(fg)} & \downarrow{\scriptstyle T^*g} \\
& & T_p^*L
\end{array}
$$

commutes.

Proof.   There is only one way to compose the ingredients, and the lemma follows since composition is associative: $\psi(fg) = (\psi f)g$. ∎

Exercise 3.4.3 | Prove that, if $\bar{f}$ is an invertible germ, then T^*f is an isomorphism.

Note 3.4.4 In the classical literature there is frequently some magic about "contravariant and covariant tensors" transforming this or that way. To some of us this is impossible to remember, but it *is* possible to remember whether our construction turns arrows around or not.

The tangent space "keeps the direction": a germ $\bar{f}\colon (M, p) \to (N, q)$ gives a linear map: $Tf\colon T_pM \to T_qN$, and the chain rule tells us that composition is OK – $T(fg) = TfTg$. The cotangent construction "turns around" the arrows: we get a map $T^*f\colon T_q^*N \to T_p^*M$ and the "cochain rule" Lemma 3.4.2 says that composition follows suit – $T^*(fg) = T^*gT^*f$.

Definition 3.4.5 The linear map $d\colon \mathcal{O}_{M,p} \to T_p^*(M)$ given by sending $\bar{\phi}\colon M \to \mathbf{R}$ to the class $d\phi \in J_p/J_p^2$ represented by $\bar{\phi} - \phi(p) = [q \mapsto \phi(q) - \phi(p)] \in J_p$ is called the *differential*.

The differential is obviously a surjection, and, when we pick an arbitrary element from the cotangent space, it is often convenient to let it be on the form $d\phi$. We note that

$$T^*f(d\phi) = d(\phi f).$$

Exercise 3.4.6 | The differential $d\colon \mathcal{O}_{M,p} \to T_p^*(M)$ is natural in the sense that, given a smooth germ $\bar{f}\colon (M, p) \to (N, q)$, the diagram

$$
\begin{array}{ccc}
\mathcal{O}_{N,q} & \xrightarrow{\;f^*\;} & \mathcal{O}_{M,p} \\
{\scriptstyle d}\big\downarrow & & {\scriptstyle d}\big\downarrow \\
T_q^*N & \xrightarrow{\;T^*f\;} & T_p^*M
\end{array}
$$

(where $f^*(\bar{\phi}) = \bar{\phi}\bar{f}$) commutes.

Lemma 3.4.7 *The differential $d\colon \mathcal{O}_{M,p} \to T_p^*(M)$ is a derivation, i.e., it is a linear map of real vector spaces satisfying the* Leibniz rule*:*

$$d(\phi \cdot \psi) = d\phi \cdot \psi + \phi \cdot d\psi,$$

where $\phi \cdot d\psi = d\psi \cdot \phi$ is the cotangent vector represented by $q \mapsto \phi(q) \cdot (\psi(q) - \psi(p))$.

Proof. We want to show that $d\phi \cdot \psi(p) + \phi(p) \cdot d\psi - d(\phi \cdot \psi)$ vanishes. It is represented by $(\bar{\phi} - \phi(p)) \cdot \bar{\psi} + \bar{\phi} \cdot (\bar{\psi} - \psi(p)) - (\bar{\phi} \cdot \bar{\psi} - \phi(p) \cdot \psi(p)) \in J_p$, which, upon collecting terms, is equal to $(\bar{\phi} - \phi(p)) \cdot (\bar{\psi} - \psi(p)) \in J_p^2$, and hence represents zero in $T_p^*M = J_p/J_p^2$. ∎

In order to relate the tangent and cotangent spaces, we need to understand the situation $(\mathbf{R}^n, 0)$. The corollary of the following lemma pins down the rôle of $J^2_{\mathbf{R}^n,0}$.

Lemma 3.4.8 *Let $\phi: U \to \mathbf{R}$ be a smooth map, where U is an open ball in $\mathbf{R}^n$ containing the origin. Then*

$$\phi(p) = \phi(0) + \sum_{i=1}^{n} p_i \cdot \phi_i(p), \qquad where \qquad \phi_i(p) = \int_0^1 D_i\phi(t \cdot p)dt.$$

Note that $\phi_i(0) = D_i\phi(0)$.

Proof. For $p \in U$ and $t \in [0, 1]$, let $F(t) = \phi(t \cdot p)$. Then $\phi(p) - \phi(0) = F(1) - F(0) = \int_0^1 F'(t)dt$ by the fundamental theorem of calculus, and $F'(t) = \sum_{i=1}^{n} p_i D_i\phi(t \cdot p)$ by the flat chain rule. ∎

Corollary 3.4.9 *The map $J_{\mathbf{R}^n,0} \to M_{1 \times n}(\mathbf{R})$ sending $\bar{\phi}$ to $D\phi(0)$ has kernel $J^2_{\mathbf{R}^n,0}$.*

Proof. The Leibniz rule implies that $J^2_{\mathbf{R}^n,0}$ is in the kernel $\{\bar{\phi} \in J_{\mathbf{R}^n,0} | D\phi(0) = 0\}$: if $\phi(p) = \psi(p) = 0$, then $D(\phi \cdot \psi)(0) = \phi(0) \cdot D\psi(0) + D\phi(0) \cdot \psi(0) = 0$. Conversely, assuming that $\phi(0) = 0$ and $D\phi(0) = 0$, the decomposition $\phi = 0 + \sum_{j=1}^{n} \mathrm{pr}_j\phi_j$ of Lemma 3.4.8 (where $\mathrm{pr}_j: \mathbf{R}^n \to \mathbf{R}$ is the jth projection, which obviously gives an element in $J_{\mathbf{R}^n,0}$) expresses $\bar{\phi}$ as an element of $J^2_{\mathbf{R}^n,0}$, since $\phi_j(0) = D_j\phi(0) = 0$. ∎

Definition 3.4.10 Let V be a real vector space. The *dual* of V, written V^*, is the vector space $\mathrm{Hom}_{\mathbf{R}}(V, \mathbf{R})$ of all linear maps $V \to \mathbf{R}$. Addition and multiplication by scalars are performed pointwise, in the sense that, if $a, b \in \mathbf{R}$ and $f, g \in V^*$, then $af + bg$ is the linear map sending $v \in V$ to $af(v) + bg(v) \in \mathbf{R}$.

If $f: V \to W$ is linear, then the *dual* linear map $f^*: W^* \to V^*$ is defined by sending $h: W \to \mathbf{R}$ to the composite $hf: V \to W \to \mathbf{R}$.

Notice that $(gf)^* = f^*g^*$.

Example 3.4.11 If $V = \mathbf{R}^n$, then any linear transformation $V \to \mathbf{R}$ is uniquely represented by a $1 \times n$ matrix, and we get an isomorphism

$$(\mathbf{R}^n)^* \cong M_{1 \times n}(\mathbf{R}) = \{v^{\mathrm{T}} \mid v \in \mathbf{R}^n\}.$$

If $f: \mathbf{R}^n \to \mathbf{R}^m$ is represented by the $m \times n$ matrix A, then $f^*: (\mathbf{R}^m)^* \to (\mathbf{R})^*$ is represented by the transpose A^{T} of A in the sense that, if $h \in (\mathbf{R}^m)^*$ corresponds to v^{T}, then $f^*(h) = hf \in (\mathbf{R}^n)^*$ corresponds to $v^{\mathrm{T}}A = (A^{\mathrm{T}}v)^{\mathrm{T}}$.

This means that, if V is a finite-dimensional vector space, then V and V^* are isomorphic (they have the same dimension), but there is no *preferred* choice of isomorphism.

Definition 3.4.12 If $\{v_1, \ldots, v_n\}$ is a basis for the vector space V, then the *dual basis* $\{v_1^*, \ldots, v_n^*\}$ for V^* is given by $v_j^* \left(\sum_{i=1}^n a_i v_i \right) = a_j$.

Exercise 3.4.13 Check that the dual basis **is** a basis and that, if $f : V \to W$ is a linear map, then the dual linear map $f^* : W^* \to V^*$ is a linear map with associated matrix the transpose of the matrix of f.

The promised natural isomorphism between the cotangent space and the dual of the tangent space is given by the following proposition.

Proposition 3.4.14 *Consider the assignment*

$$\alpha = \alpha_{M,p} : T_p^* M \to (T_p M)^*, \qquad d\phi \mapsto \{[\gamma] \mapsto (\phi\gamma)'(0)\}.$$

1. *$\alpha_{M,p}$ is a well-defined linear map.*
2. *$\alpha_{M,p}$ is natural in (M, p), in the sense that if $\bar{f} : (M, p) \to (N, q)$ is a germ, then the diagram*

$$
\begin{array}{ccc}
T_q^* N & \xrightarrow{\ T^* f\ } & T_p^* M \\
{\scriptstyle \alpha_{N,q}} \downarrow & & \downarrow {\scriptstyle \alpha_{M,p}} \\
(T_q N)^* & \xrightarrow{\ (Tf)^*\ } & (T_p M)^*
\end{array}
$$

commutes.
3. *Let (x, U) be a chart for M with $p \in U$, and $A_x : T_p M \to \mathbf{R}^m$ the isomorphism of Lemma 3.3.10 given by $A_x[\gamma] = (x\gamma)'(0)$. Then the composite*

$$T_p^* M \xrightarrow{\ \alpha_{M,p}\ } (T_p M)^* \xrightarrow[\cong]{\ (A_x^{-1})^*\ } (\mathbf{R}^m)^*$$

sends the cotangent vector $d\phi$ to (the linear transformation $\mathbf{R}^m \to \mathbf{R}$ given by multiplication with) the Jacobi matrix $D(\phi x^{-1})(x(p))$. In particular, under this isomorphism, dx_i correspond to $e_i^* = e_i^T \cdot$, the dual of the ith standard basis vector in $\mathbf{R}^n$.
4. *$\alpha_{M,p}$ is an isomorphism.*

Proof.

1. We show that J_p^2 is in the kernel of the (well-defined) linear transformation

$$J_p \to (T_p M)^*, \qquad \bar{\phi} \mapsto \{[\gamma] \mapsto (\phi\gamma)'(0)\}.$$

If $\phi(p) = \psi(p) = 0$, then the Leibniz rule gives

$$((\phi \cdot \psi)\gamma)'(0) = ((\phi\gamma) \cdot (\psi\gamma))'(0) = (\phi\gamma)(0) \cdot (\psi\gamma)'(0) + (\phi\gamma)'(0) \cdot (\psi\gamma)(0) = 0,$$

regardless of $[\gamma] \in T_p M$.

2. Write out the definitions and conclude that both ways around the square send a cotangent vector $d\phi$ to the linear map $\{[\gamma] \mapsto (\phi \circ f \circ \gamma)'(0)\}$.

3. Recalling that $A_x^{-1}(v) = [t \mapsto x^{-1}(x(p)+tv)]$ we get that the composite sends the cotangent vector $d\phi$ to the element in $(\mathbf{R}^m)^*$ given by sending $v \in \mathbf{R}^m$ to the derivative at 0 of $t \mapsto \phi x^{-1}(x(p) + tv)$, which by the chain rule is exactly $D(\phi x^{-1})(x(p)) \cdot v$.

4. By naturality, we just have to consider the case $(M, p) = (\mathbf{R}^m, 0)$ (use naturality as in Exercises 3.3.8 and 3.4.3 with $\bar{f}$ the germ of a chart). Hence we are reduced to showing that the composite $(A_x^{-1})^* \alpha_{\mathbf{R}^m,0}$ is an isomorphism when x is the identity chart. But this is exactly Corollary 3.4.9: the kernel of $J_{\mathbf{R}^m,0} \to (\mathbf{R}^m)^* \cong M_{1 \times m}(\mathbf{R})$ sending $\bar{\phi}$ to $D\phi(0)$ is precisely $J_{\mathbf{R}^m,0}^2$ and so the induced map from $T_0^*\mathbf{R}^m = J_{\mathbf{R}^m,0}/J_{\mathbf{R}^m,0}^2$ is an isomorphism. ∎

In order to get a concrete grip on the cotangent space, we should understand the linear algebra of dual vector spaces a bit better.

Note 3.4.15 Take (x, U) as a chart for M around $p \in M$ and let $x_i = \mathrm{pr}_i x$ be the "ith coordinate". The isomorphism $(A_x^{-1})^* \alpha_{M,p} \colon T_p^* M \overset{\cong}{\to} (\mathbf{R}^m)^*$ of Proposition 3.4.14(3) sends dx_i to multiplication by the Jacobi matrix $D(x_i x^{-1})(x(p)) = D(\mathrm{pr}_i)(x(p)) = [0\,0\ldots0\,1\,0\ldots0\,0] = e_i^{\mathrm{T}}$, i.e., the dual of the ith standard basis vector e_i:

$$T_p^* M \xrightarrow[\cong]{dx_i \leftrightarrow e_i^*} (\mathbf{R}^m)^*.$$

Consequently, $\{dx_i\}_{i=1,\ldots,m}$ is a basis for the cotangent space $T_p^* M$.

Exercise 3.4.16	Verify the claim in Note 3.4.15. Also show that

$$d\phi = \sum_{i=1}^{n} D_i(\phi x^{-1})(x(p)) \cdot dx_i.$$

To get notation as close as possible to the classical notation, one often writes $\partial\phi/\partial x_i(p)$ instead of $D_i(\phi x^{-1})(x(p))$, and gets the more familiar expression

$$d\phi = \sum_{i=1}^{n} \frac{\partial\phi}{\partial x_i}(p) \cdot dx_i.$$

One good thing about understanding manifolds is that we finally can answer the following questions. "What is the x in that formula. What is actually meant by 'variables', and what is the mysterious symbol 'dx_i'?" Here x is the name of a particular chart. In the special case where $x = \mathrm{id}\colon \mathbf{R}^n = \mathbf{R}^n$ we see that x_i is just a name for the projection onto the ith coordinate and $D_i(\phi)(p) = \partial\phi/\partial x_i(p)$.

Hence, the Jacobian of a smooth function $f: M \to \mathbf{R}$ in some chart (x, U) is nothing but the vector corresponding to df in the basis $\{dx_i\}_{i=1,\dots,n}$.

In view of the isomorphism $T_p^*\mathbf{R}^n \cong \mathbf{R}^n$, $d\phi \mapsto [D_1\phi(p), \dots, D_n\phi(p)]^T$ some authors refer to the differential $d\phi$ as the *gradient* of ϕ. We will resist this temptation for reasons that will become clear when we talk about Riemannian structures: even though T_pM (where the gradient should live) and T_p^*M are abstractly isomorphic, we need a *choice* of isomorphism and for general M we will want to control how this choice varies with p.

Example 3.4.17 In the robot example in Section 1.1, we considered a function $f: S^1 \times S^1 \to \mathbf{R}^1$ given by

$$f(e^{i\theta}, e^{i\phi}) = |3 - e^{i\theta} - e^{i\phi}| = \sqrt{11 - 6\cos\theta - 6\cos\phi + 2\cos(\theta - \phi)},$$

and so, expressed in the basis of the angle charts as in Exercise 2.2.12, the differential is

$$df(e^{i\theta}, e^{i\phi})$$
$$= \frac{(3\sin\theta - \cos\phi\sin\theta + \sin\phi\cos\theta)d\theta + (3\sin\phi - \cos\theta\sin\phi + \sin\theta\cos\phi)d\phi}{f(e^{i\theta}, e^{i\phi})}.$$

Note 3.4.18 Via the correspondence between a basis and its dual in terms of transposition we can explain the classical language of "transforming this or that way". If $x: \mathbf{R}^n \cong \mathbf{R}^n$ is a diffeomorphism (and hence is a chart in the standard smooth structure of $\mathbf{R}^n$, or a "change of coordinates") and $p \in \mathbf{R}^n$, then the diagram

$$
\begin{array}{ccc}
T_p\mathbf{R}^n & \xrightarrow{\ T_px\ } & T_{x(p)}\mathbf{R}^n \\
{\scriptstyle [\gamma]\mapsto\gamma'(0)}\downarrow{\scriptstyle\cong} & & {\scriptstyle [\gamma]\mapsto\gamma'(0)}\downarrow{\scriptstyle\cong} \\
\mathbf{R}^n & \xrightarrow{\ Dx(p)\cdot\ } & \mathbf{R}^n
\end{array}
$$

commutes, that is, the change of coordinates x transforms tangent vectors by multiplication by the Jacobi matrix $Dx(p)$. For cotangent vectors the situation is that

$$
\begin{array}{ccc}
T_{x(p)}^*\mathbf{R}^n & \xrightarrow{\ T^*x\ } & T_p^*\mathbf{R}^n \\
{\scriptstyle\cong}\downarrow{\scriptstyle d\phi\mapsto[D\phi(x(p))]^T} & & {\scriptstyle\cong}\downarrow{\scriptstyle d\phi\mapsto[D\phi(p)]^T} \\
\mathbf{R}^n & \xrightarrow{\ [Dx(p)]^T\cdot\ } & \mathbf{R}^n
\end{array}
$$

commutes.

Exercise 3.4.19   Let $0 \neq p \in M = \mathbf{R}^2 = \mathbf{C}$, let $x \colon \mathbf{R}^2 = \mathbf{R}^2$ be the identity chart and let $y \colon V \cong V'$ be polar coordinates: $y^{-1}(r, \theta) = re^{i\theta}$, where V is $\mathbf{C}$ minus some ray from the origin not containing p, and V' the corresponding strip of radii and angles. Show that the upper horizontal arrow in

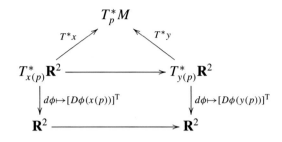

is $T^*(xy^{-1})$ and the lower horizontal map is given by multiplication by the transposed Jacobi matrix $D(xy^{-1})(y(p))^{\mathrm{T}}$, and calculate this explicitly in terms of p_1 and p_2.

Conversely, in the same diagram with tangent spaces instead of cotangent spaces (remove the superscript *, reverse the diagonal maps, and let the vertical maps be given by $[\gamma] \mapsto (x\gamma)'(0)$ and $[\gamma] \mapsto (y\gamma)'(0)$ respectively), show that the upper horizontal map is $T_{x(p)}(yx^{-1})$ and the lower one is given by multiplication with the Jacobi matrix $D(yx^{-1})(x(p))$, and calculate this explicitly in terms of p_1 and p_2.

Example 3.4.20   If this example makes no sense to you, don't worry, it's for the physicists among us! Classical mechanics is all about the relationship between the tangent space and the cotangent space. More precisely, the kinetic energy E should be thought of as (half) an *inner product* g on the tangent space, i.e., as a symmetric bilinear and positive definite map

$$g = 2E \colon T_pM \times T_pM \to \mathbf{R}.$$

This is the equation $E = \frac{1}{2}m|v|^2$ you know from high school, giving the kinetic energy as something proportional to the norm applied to the velocity v. The usual – mass-independent – inner product in Euclidean space gives $g(v, v) = v^{\mathrm{T}} \cdot v = |v|^2$, in mechanics the mass is incorporated into the inner product.

The assignment $[\gamma] \mapsto g([\gamma], -)$, where $g([\gamma], -) \colon T_pM \to \mathbf{R}$ is the linear map $[\gamma_1] \mapsto g([\gamma], [\gamma_1])$, defines an isomorphism $T_pM \cong (T_pM)^* = \mathrm{Hom}_{\mathbf{R}}(T_pM, \mathbf{R})$ (isomorphism since g is positive definite). The *momentum* of a particle with mass m moving along the curve γ is, at time $t = 0$, exactly the cotangent vector $g([\gamma], -)$ (this is again the old formula $p = mv$: the mass is intrinsic to the inner product, and the v should really be transposed ($p = g(v, -) = mv^{\mathrm{T}}$) so as to be ready to be multiplied by another v to give $E = \frac{1}{2}m|v|^2 = \frac{1}{2}p \cdot v$).

3.5 Derivations[1]

Although the definition of the tangent space by means of curves is very intuitive and geometric, the alternative point of view of the tangent space as the space of "derivations" can be very convenient. A derivation is a linear transformation satisfying the Leibniz rule.

Definition 3.5.1 Let M be a smooth manifold and $p \in M$. A *derivation* (on M at p) is a linear transformation

$$X: \mathcal{O}_{M,p} \to \mathbf{R}$$

satisfying the *Leibniz rule*

$$X(\bar{\phi} \cdot \bar{\psi}) = X(\bar{\phi}) \cdot \psi(p) + \phi(p) \cdot X(\bar{\psi})$$

for all function germs $\bar{\phi}, \bar{\psi} \in \mathcal{O}_{M,p}$.
   We let $D|_p M$ be the set of all derivations.

Example 3.5.2 Let $M = \mathbf{R}$. Then $\bar{\phi} \mapsto \phi'(p)$ is a derivation. More generally, if $M = \mathbf{R}^n$ then all the partial derivatives $\bar{\phi} \mapsto D_j(\phi)(p)$ are derivations.

Note 3.5.3 Note that the set $D|_p M$ of derivations is a vector space: adding two derivations or multiplying one by a real number gives a new derivation. We shall later see that the partial derivatives form a basis for the vector space $D|_p \mathbf{R}^n$.

Definition 3.5.4 Let $\bar{f}: (M, p) \to (N, f(p))$ be a germ. Then we define the linear transformation

$$D|_p f: D|_p M \to D|_{f(p)} N$$

by

$$D|_p f(X) = X f^*$$

(i.e. $D|_p f(X)(\bar{\phi}) = X(\phi f)$).

Lemma 3.5.5 *If $\bar{f}: (M, p) \to (N, f(p))$ and $\bar{g}: (N, f(p)) \to (L, g(f(p)))$ are germs, then*

$$
\begin{array}{ccc}
D|_p M & \xrightarrow{D|_p f} & D|_{f(p)} N \\
& \searrow{\scriptstyle D|_p(gf)} & \downarrow{\scriptstyle D|_{f(p)}g} \\
& & D|_{gf(p)} L
\end{array}
$$

commutes.

Exercise 3.5.6 Prove Lemma 3.5.5.

[1] This material is not used in an essential way in the rest of the book. It is included for completeness, and for comparison with other sources.

3.5.1 The Space of Derivations Is the Dual of the Cotangent Space

Given our discussion of the cotangent space $T_p^* M = J_p/J_p^2$ in the previous section, it is easy to identify the space of derivations as the dual of the cotangent space (and hence the *double dual* of the tangent space).[2]

However, it is instructive to see how naturally the derivations fall out of our discussion of the cotangent space (this is of course a reflection of a deeper theory of derivations you may meet later if you study algebra).

Proposition 3.5.7 *Let M be a smooth manifold and $p \in M$. Then*

$$\beta_{M,p} \colon \left(T_p^* M\right)^* \longrightarrow D|_p M, \qquad \beta_{M,p}(g) = \{\mathcal{O}_{M,p} \xrightarrow{\bar{\phi} \mapsto d\phi} T_p^* M \xrightarrow{g} \mathbf{R}\}$$

is a natural isomorphism; if $f \colon M \to N$ is a smooth map, then

$$
\begin{array}{ccc}
\left(T_p^* M\right)^* & \xrightarrow{\ \beta_{M,p}\ } & D|_p M \\
{\scriptstyle (T^* f)^*} \downarrow & & \downarrow {\scriptstyle D|_p f} \\
\left(T_{f(p)}^* N\right)^* & \xrightarrow{\ \beta_{N,f(p)}\ } & D|_{f(p)} N
\end{array}
$$

commutes.

Proof. Recall that $T_p^* M = J_p/J_p^2$, where $J_p \subseteq \mathcal{O}_{M,p}$ consists of the germs vanishing at p. That $\beta_{M,p}(g)$ is a derivation follows since g is linear and since d satisfies the Leibniz rule by Lemma 3.4.7: $\beta_{M,p}(g)$ applied to $\bar{\phi} \cdot \bar{\psi}$ gives $g(d(\phi \cdot \psi)) = \phi(p) \cdot g(d\psi) + g(d\phi) \cdot \psi(p)$. The inverse of $\beta_{M,p}$ is given as follows. Given a derivation $h \colon \mathcal{O}_{M,p} \to \mathbf{R}$, notice that the Leibniz rule gives that $J_p^2 \subseteq \ker\{h\}$, and so h defines a map $\beta_{M,p}^{-1}(h) \colon J_p/J_p^2 \to \mathbf{R}$.

Showing that the diagram commutes boils down to following an element $g \in (T_p^* M)^*$ both ways and observing that the result either way is the derivation sending $\bar{\phi} \in \mathcal{O}_{M,p}$ to $g(d(\phi f)) \in \mathbf{R}$. ∎

For a vector space V, there is a canonical map to the double dualization $V \to (V^*)^*$ sending $v \in V$ to $v^{**} \colon V^* \to \mathbf{R}$ given by $v^{**}(f) = f(v)$. This map is always injective, and if V is finite-dimensional it is an isomorphism. This is also natural: if $f \colon V \to W$ is linear, then

$$
\begin{array}{ccc}
V & \longrightarrow & (V^*)^* \\
{\scriptstyle f} \downarrow & & \downarrow {\scriptstyle (f^*)^*} \\
W & \longrightarrow & (W^*)^*
\end{array}
$$

commutes.

---

[2] For the benefit of those who did not study the cotangent space, we give an independent proof of this fact in the next subsection, along with some further details about the structure of the space of derivations.

Together with the above result, this gives the promised natural isomorphism between the double dual of the tangent space and the space of derivations:

Corollary 3.5.8 *There is a chain of natural isomorphism*

$$T_pM \xrightarrow{\;\cong\;} ((T_pM)^*)^* \xrightarrow{(\alpha_{M,p})^*} (T_p^*M)^* \xrightarrow{\beta_{M,p}} D|_pM.$$

The composite sends $[\gamma] \in T_pM$ to $X_\gamma \in D|_pM$ whose value at $\bar{\phi} \in \mathcal{O}_{M,p}$ is $X_\gamma(\bar{\phi}) = (\phi\gamma)'(0)$.

Note 3.5.9   In the end, this all sums up to say that T_pM and $D|_pM$ are one and the same thing (the categorists would say that "the functors are naturally isomorphic"), and so we will let the notation $D|_pM$ slip quietly into oblivion.

Notice that in the proof of Corollary 3.5.8 it is crucial that the tangent spaces are finite-dimensional. However, the proof of Proposition 3.5.7 is totally algebraic, and does not depend on finite dimensionality.

3.5.2 The Space of Derivations Is Spanned by Partial Derivatives

Even if we know that the space of derivations is just another name for the tangent space, a bit of hands-on knowledge about derivations can often be useful. This subsection does not depend on the previous one, and as a side effect gives a direct proof of $T_pM \cong D|_pM$ without talking about the cotangent space.

The chain rule gives, as before, that we may use charts and transport all calculations to $\mathbf{R}^n$.

Proposition 3.5.10   *The partial derivatives $\{D_i|_0\}i = 1, \dots, n$ form a basis for $D|_0\mathbf{R}^n$.*

Exercise 3.5.11   Prove Proposition 3.5.10.

Thus, given a chart $\bar{x}: (M, p) \to (\mathbf{R}^n, 0)$ we have a basis for $D|_pM$, and we give this basis the old-fashioned notation to please everybody.

Definition 3.5.12   Consider a chart $\bar{x}: (M, p) \to (\mathbf{R}^n, x(p))$. Define the derivation in T_pM

$$\left.\frac{\partial}{\partial x_i}\right|_p = (D|_px)^{-1}(D_i|_{x(p)}),$$

or, in more concrete language, if $\bar{\phi}: (M, p) \to (\mathbf{R}, \phi(p))$ is a function germ, then

$$\left.\frac{\partial}{\partial x_i}\right|_p(\bar{\phi}) = D_i(\phi x^{-1})(x(p)).$$

Note 3.5.13 Note that, if $\bar{f} \colon (M, p) \to (N, f(p))$ is a germ, then the matrix associated with the linear transformation $D|_p f \colon D|_p M \to D|_{f(p)} N$ in the basis given by the partial derivatives of x and y is nothing but the Jacobi matrix $D(yfx^{-1})(x(p))$. In the current notation the i, j entry is

$$\left. \frac{\partial (y_i f)}{\partial x_j} \right|_p .$$

Definition 3.5.14 Let M be a smooth manifold and $p \in M$. With every germ $\bar{\gamma} \colon (\mathbf{R}, 0) \to (M, p)$ we may associate a derivation $X_\gamma \colon \mathcal{O}_{M,p} \to \mathbf{R}$ by setting

$$X_\gamma(\bar{\phi}) = (\phi\gamma)'(0)$$

for every function germ $\bar{\phi} \colon (M, p) \to (\mathbf{R}, \phi(p))$.

Note that $X_\gamma(\bar{\phi})$ is the derivative at zero of the composite

$$(\mathbf{R}, 0) \xrightarrow{\ \bar{\gamma}\ } (M, p) \xrightarrow{\ \bar{\phi}\ } (\mathbf{R}, \phi(p)).$$

Exercise 3.5.15 Check that the map $T_p M \to D|_p M$ sending $[\gamma]$ to X_γ is well defined.

Using the definitions we get the following lemma, which says that the map $T_0 \mathbf{R}^n \to D|_0 \mathbf{R}^n$ is surjective.

Lemma 3.5.16 *If $v \in \mathbf{R}^n$ and $\bar{\gamma}$ is the germ associated with the curve $\gamma(t) = v \cdot t$, then*

$$X_\gamma(\bar{\phi}) = D(\phi)(0) \cdot v = \sum_{i=0}^{n} v_i D_i(\phi)(0).$$

In particular, if $v = e_j$ is the jth unit vector, then $X_\gamma = D_j$ is the jth partial derivative at zero.

Lemma 3.5.17 *Let $\bar{f} \colon (M, p) \to (N, f(p))$ be a germ. Then*

$$
\begin{array}{ccc}
T_p M & \xrightarrow{\ T_p f\ } & T_{f(p)} N \\
\downarrow & & \downarrow \\
D|_p M & \xrightarrow{\ D|_p f\ } & D|_{f(p)} N
\end{array}
$$

commutes.

Exercise 3.5.18 Prove Lemma 3.5.17.

Proposition 3.5.19 *Let M be a smooth manifold and p a point in M. The assignment $[\gamma] \mapsto X_\gamma$ defines a natural isomorphism*

$$T_p M \cong D|_p M$$

between the tangent space at p and the vector space of derivations $\mathcal{O}_{M,p} \to \mathbf{R}$.

Proof. The term "natural" in the proposition refers to the statement in Lemma 3.5.17. In fact, we can use this to prove the rest of the proposition.

Choose a germ chart $\bar{x} \colon (M, p) \to (\mathbf{R}^n, 0)$. Then Lemma 3.5.17 proves that

$$
\begin{array}{ccc}
T_p M & \xrightarrow[\cong]{T_p x} & T_0 \mathbf{R}^n \\
\downarrow & & \downarrow \\
D|_p M & \xrightarrow[\cong]{D|_p x} & D|_0 \mathbf{R}^n
\end{array}
$$

commutes, and the proposition follows if we know that the right-hand map is a linear isomorphism.

But we have seen in Proposition 3.5.10 that $D|_0 \mathbf{R}^n$ has a basis consisting of partial derivatives, and we noted in Lemma 3.5.16 that the map $T_0 \mathbf{R}^n \to D|_0 \mathbf{R}^n$ hits all the basis elements, and now the proposition follows since the dimension of $T_0 \mathbf{R}^n$ is n (a surjective linear map between vector spaces of the same (finite) dimension is an isomorphism). ∎

4 Regular Values

In this chapter we will acquire a powerful tool for constructing new manifolds as inverse images of smooth functions. This result is a consequence of the *rank theorem*, which says roughly that smooth maps are – locally around "most" points – like linear projections or inclusions of Euclidean spaces.

4.1 The Rank

Remember that the rank of a linear transformation is the dimension of its image. In terms of matrices, this can be captured by saying that a matrix has rank at least r if it contains an $r \times r$ invertible submatrix.

Definition 4.1.1 Let $\bar{f} \colon (M, p) \to (N, f(p))$ be a smooth germ. The *rank* $\operatorname{rk}_p f$ *of* f *at* p is the rank of the linear map $T_p f$. We say that a germ $\bar{f}$ has *constant rank* r if it has a representative $f \colon U_f \to N$ whose rank $\operatorname{rk} T_q f = r$ for all $q \in U_f$. We say that a germ $\bar{f}$ has rank $\geq r$ if it has a representative $f \colon U_f \to N$ whose rank $\operatorname{rk} T_q f \geq r$ for all $q \in U_f$.

In view of Proposition 3.3.11, the rank of f at p is the same as the rank of the Jacobi matrix $D(yfx^{-1})(x(p))$, where (x, U) is a chart around p and (y, V) a chart around $f(p)$.

Lemma 4.1.2 *Let $\bar{f} \colon (M, p) \to (N, f(p))$ be a smooth germ. If $\operatorname{rk}_p f = r$ then there exists an open neighborhood U of p such that $\operatorname{rk}_q f \geq r$ for all $q \in U$.*

Proof. Note that the subspace $M_{n \times m}^{\geq r}(\mathbf{R}) \subseteq M_{n \times m}(\mathbf{R})$ of $n \times m$ matrices of rank at least r is open: the determinant function is continuous, so the set of matrices such that a given $r \times r$ submatrix is invertible is open (in fact, if for $S \subseteq \{1, \ldots n\}$ and $T \subseteq \{1, \ldots, m\}$ being two sets with r elements each we let $\det_{S,T} \colon M_{n \times m}(\mathbf{R}) \to \mathbf{R}$ be the continuous function sending the $n \times m$ matrix (a_{ij}) to $\det((a_{ij})_{i \in S, j \in T})$ we see that $M_{n \times m}^{\geq r}(\mathbf{R})$ is the finite union $\bigcup_{S,T} \det_{S,T}^{-1}(\mathbf{R} \setminus \{0\})$ of open sets).

Choose a representative $f \colon U_f \to N$ and charts (x, U) and (y, V) with $p \in U$ and $f(p) \in V$. Let $W = U_f \cap U \cap f^{-1}(V)$, and consider the continuous function $J \colon W \to M_{n \times m}(\mathbf{R})$ sending $q \in W$ to the Jacobian $J(q) = D(yfx^{-1})(x(q))$. The desired neighborhood of p is then $J^{-1}(M_{n \times m}^{\geq r}(\mathbf{R}))$. ∎

Note 4.1.3 In the previous proof we used the useful fact that the subspace $M_{n \times m}^{\geq r}(\mathbf{R}) \subseteq M_{n \times m}(\mathbf{R})$ of $n \times m$ matrices of rank at least r is open. As a matter of fact, we showed in Example 2.5.10 that the subspace $M_{n \times m}^{r}(\mathbf{R}) \subseteq M_{n \times m}(\mathbf{R})$ of $n \times m$ matrices of rank (exactly equal to) r is a submanifold of codimension $(m - r)(n - r)$. Perturbing a rank-r matrix may kick you out of this manifold and into one of higher rank (but if the perturbation is small enough you can avoid the matrices of smaller rank).

To remember what way the inequality in Lemma 4.1.2 goes, it may help to recall that the zero matrix is the only matrix of rank 0 (and so all the neighboring matrices are of higher rank), and likewise that the subset $M_{n \times m}^{\min(m,n)}(\mathbf{R}) \subseteq M_{n \times m}(\mathbf{R})$ of matrices of maximal rank is open. The rank "does not decrease locally".

Example 4.1.4 The map $f : \mathbf{R} \to \mathbf{R}$ given by $f(p) = p^2$ has $Df(p) = 2p$, and so

$$
\mathrm{rk}_p f = \begin{cases} 0 & p = 0 \\ 1 & p \neq 0. \end{cases}
$$

Exercise 4.1.5 What is the rank of the function $f : \mathbf{R}^2 \to \mathbf{R}^2$ given by $f(s, t) = (s^2, st)$?

Example 4.1.6 Consider the determinant $\det : M_2(\mathbf{R}) \to \mathbf{R}$ with

$$
\det(A) = a_{11}a_{22} - a_{12}a_{21}, \text{ for } A = \begin{bmatrix} a_{11} & a_{12} \\ a_{21} & a_{22} \end{bmatrix}.
$$

By the calculation in Example 3.3.15 we see that

$$
\mathrm{rk}_A \det = \begin{cases} 0 & A = 0 \\ 1 & A \neq 0. \end{cases}
$$

For dimension $n \geq 2$, the analogous statement is that $\mathrm{rk}_A \det = 0$ if and only if $\mathrm{rk}\, A < n - 1$.

Example 4.1.7 Consider the map $f : S^1 \subseteq \mathbf{C} \to \mathbf{R}$ given by $f(x+iy) = x$. Cover S^1 by the angle charts $x : S^1 - \{1\} \to (0, 2\pi)$ and $y : S^1 - \{-1\} \to (-\pi, \pi)$ with $x^{-1}(t) = y^{-1}(t) = e^{it}$ (whenever defined, c.f. Exercise 2.2.12). Then $fx^{-1}(t) = fy^{-1}(t) = \cos(t)$, and so we see that the rank of f at z is 1 if $z \neq \pm 1$ and 0 if $z = \pm 1$.

Definition 4.1.8 Let $f : M \to N$ be a smooth map where N is n-dimensional. A point $p \in M$ is *regular* if $T_p f$ is surjective (i.e., if $\mathrm{rk}_p f = n$). A point $q \in N$ is a *regular value* if all $p \in f^{-1}(q)$ are regular points. Synonyms for "non-regular" are *critical* or *singular*.

Note that a point q which is not in the image of f is a regular value since $f^{-1}(q) = \emptyset$.

Note 4.1.9 These names are well chosen: the critical values are critical in the sense that they exhibit bad behavior. The inverse image $f^{-1}(q) \subseteq M$ of a regular value q will turn out to be a submanifold, whereas inverse images of critical points usually are not.

On the other hand, according to Sard's theorem (Theorem 4.6.1) the regular values are the common state of affairs (in technical language: critical values have "measure zero" while regular values are "dense").

Example 4.1.10 The names correspond to the normal usage in multivariable calculus. For instance, if you consider the function

$$f : \mathbf{R}^2 \to \mathbf{R}$$

whose graph is depicted in Figure 4.1, the critical points – i.e., the points $p \in \mathbf{R}^2$ such that

$$D_1 f(p) = D_2 f(p) = 0$$

– will correspond to the two local maxima and the saddle point. We note that the contour lines at all other values are nice one-dimensional submanifolds of $\mathbf{R}^2$ (circles, or disjoint unions of circles).

In Figure 4.2, we have considered a standing torus, and looked at its height function. The contour lines are then inverse images of various height values. If we had written out the formulae we could have calculated the rank of the height function at every point of the torus, and we would have found four critical points: one on the top, one on "the top of the hole", one on "the bottom of the hole" (the point on the figure where you see two contour lines cross) and one on the bottom. The contours at these heights look like points or figure eights, whereas contour lines at other values are one or two circles.

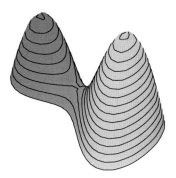

Figure 4.1.

Figure 4.2.

Example 4.1.11 In the robot example in Section 1.1, we considered a function

$$f \colon S^1 \times S^1 \to \mathbf{R}^1$$

and found three critical values. To be more precise,

$$f(e^{i\theta}, e^{i\phi}) = |3 - e^{i\theta} - e^{i\phi}| = \sqrt{11 - 6\cos\theta - 6\cos\phi + 2\cos(\theta - \phi)},$$

and so (using charts corresponding to the angles as in Exercise 2.2.12: conveniently all charts give the same formulae in this example) the Jacobi matrix at $(e^{i\theta}, e^{i\phi})$ equals

$$\frac{1}{f(e^{i\theta}, e^{i\phi})}[3\sin\theta - \cos\phi\sin\theta + \sin\phi\cos\theta, 3\sin\phi - \cos\theta\sin\phi + \sin\theta\cos\phi].$$

The rank is one, unless both coordinates are zero, in which case we get that we must have $\sin\theta = \sin\phi = 0$, which leaves the points

$$(1, 1), \qquad (-1, -1), \qquad (1, -1), \qquad (-1, 1),$$

giving the critical values 1, 5 and (twice) 3: exactly the points we noticed as troublesome.

Exercise 4.1.12 The rank of a smooth map is equal to the rank of the *cotangent* map.

Exercise 4.1.13 Fill out the details in the robot example of Section 1.1.

4.2 The Inverse Function Theorem

The technical foundation for the theorems to come is the inverse function theorem from multivariable calculus which we cite below. A proof can be found in Theorem 2.11 of [19], or in any other decent book on multivariable calculus.

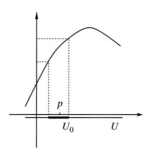

Figure 4.3.

Theorem 4.2.1 *Let $f : U_1 \to U_2$ be a smooth function where $U_1, U_2 \subseteq \mathbf{R}^n$. Let $p \in U_1$ and assume the Jacobi matrix $Df(p)$ is invertible in the point p. Then there exists a neighborhood around p on which f is smoothly invertible, i.e., there exists an open subset $U_0 \subseteq U_1$ containing p such that*

$$f|_{U_0} : U_0 \to f(U_0)$$

is a diffeomorphism onto an open subset of U_2. See Figure 4.3.

Note that the flat chain rule, Lemma 3.0.3, gives that the inverse has Jacobian matrix

$$D(f^{-1})(f(x)) = [Df(x)]^{-1}.$$

Recall from Lemma 3.1.7 that an invertible germ $(M, p) \to (N, q)$ is exactly a germ induced by a diffeomorphism $\phi \colon U \to V$ between neighborhoods of p and q.

Theorem 4.2.2 (The inverse function theorem) *A germ*

$$\bar{f} : (M, p) \to (N, f(p))$$

is invertible if and only if

$$T_p f : T_p M \to T_{f(p)} N$$

is invertible, in which case $T_{f(p)}(f^{-1}) = (T_p f)^{-1}$.

Proof. Choose charts (x, U) and (y, V) with $p \in W = U \cap f^{-1}(V)$. By Proposition 3.3.11, $T_p f$ is an isomorphism if and only if the Jacobi matrix $D(yfx^{-1})(x(p))$ is invertible (which incidentally implies that $\dim(M) = \dim(N)$).

By the inverse function theorem, 4.2.1 in the flat case, this is the case if and only if yfx^{-1} is a diffeomorphism when restricted to a neighborhood $U_0 \subseteq x(U)$ of $x(p)$. As x and y are diffeomorphisms, this is the same as saying that $f|_{x^{-1}(U_0)}$ is a diffeomorphism. ∎

The inverse function theorem has two very useful corollaries. The first follows directly from Lemma 3.1.7.

Corollary 4.2.3 *A germ* $\bar{f}: (M, p) \to (\mathbf{R}^m, q)$ *is represented by a chart* $x: U \to U'$ *if and only if* $T_p f$ *is invertible.*

Corollary 4.2.4 *Let* $f: M \to N$ *be a smooth map between smooth n-dimensional manifolds. Then* f *is a diffeomorphism if and only if it is bijective and* $T_p f$ *is of rank n for all* $p \in M$.

Proof. One way is obvious. For the other implication assume that f is bijective and $T_p f$ is of rank n for all $p \in M$. Since f is bijective it has an inverse function. A function has at most one inverse function(!), so the smooth inverse functions existing locally by virtue of the inverse function theorem must be equal to the globally defined inverse function, which hence is smooth. ∎

Exercise 4.2.5 Let G be a Lie group (a smooth manifold with a smooth associative multiplication, with a unit and all inverses). Show that the map $G \to G$ given by sending an element g to its inverse g^{-1} is smooth. (Some authors have this as a part of the definition of a Lie group, which is totally redundant. However, if G is only a topological space with a continuous associative multiplication, with a unit and all inverses, it does not automatically follow that inverting elements gives a continuous function.)

4.3 The Rank Theorem

The rank theorem says that, if the rank of a smooth map $f: M \to N$ is constant in a neighborhood of a point, then there are charts such that f looks like a composite $\mathbf{R}^m \to \mathbf{R}^r \subseteq \mathbf{R}^n$, where the first map is the projection onto the first $r \leq m$ coordinate directions, and the last one is the inclusion of the first $r \leq n$ coordinates. So, for instance, a map of rank 1 between 2-manifolds looks locally like

$$\mathbf{R}^2 \to \mathbf{R}^2, \qquad (q_1, q_2) \mapsto (q_1, 0).$$

Not only does the rank theorem have an enormous impact, but also its proof carries two very neat ideas, namely

(1) if the question is "local", we may reduce to the case where our manifolds are Euclidean spaces, and
(2) the inverse function theorem (in the guise of Corollary 4.2.3) is an extremely efficient device for checking that a concrete formula actually gives rise to a chart – you simply calculate the Jacobian and observe that it is invertible.

As we present it, the rank theorem comes in four different flavors, which can be a bit much for a first reading. At this stage, the reader might want to focus on a variant (the third, which is the one relevant for discussing regular values) and return to the full version when the other variations come into play. So, the advice is to study Lemma 4.3.1 and its proof closely, then jump directly to Exercise 4.3.4 and take in the rank theorem, 4.3.3, during the second reading.

Lemma 4.3.1 *(The Rank Theorem, Regular Value/"Third" Case) Let $\bar{f}$: $(M, p) \to (N, q)$ be a germ with $\mathrm{rk}_p f = \dim N$. Then, for any chart (y, V) around q, there exists a chart (x, U) around p such that*

$$\overline{yfx^{-1}} = \overline{\mathrm{pr}},$$

where $\mathrm{pr}\colon \mathbf{R}^{\dim M} \to \mathbf{R}^{\dim N}$ *is the projection* $\mathrm{pr}(t_1, \ldots, t_{\dim M}) = (t_1, \ldots, t_{\dim N})$.

Proof. Let $m = \dim M$ and $n = \dim N$ and let $f\colon U_f \to N$ be some representative of $\bar{f}$, where we may assume that $U_f \subseteq f^{-1}(V)$.

First we do the case where $y(q) = 0$. On choosing *any* chart (x_1, U_1) around p with $x_1(p) = 0$ we get that $Dyfx_1^{-1}(0)$ has rank n. If need be, we may permute the coordinates of x_1 (which has the effect of permuting the columns of $Dyfx_1^{-1}(0)$) so that

$$Dyfx_1^{-1}(0) = [A\ B],$$

where $A \in \mathrm{GL}_n(\mathbf{R})$. Letting $g = yfx_1^{-1}|_{x_1(U_f \cap U_1)}$ and $x_2(t) = (g(t), t_{n+1}, \ldots, t_m)$ for t in the domain of g, we get that

$$Dx_2(0) = \begin{bmatrix} A & B \\ 0 & I \end{bmatrix}.$$

This matrix is invertible, so, by Corollary 4.2.3 of the inverse function theorem, x_2 defines a diffeomorphism $x_2\colon U_2 \cong U_2'$ where U_2 and U_2' both are neighborhoods of 0 in $\mathbf{R}^m$.

For $t \in U_2'$ we have

$$\mathrm{pr}(t) = \mathrm{pr}(x_2 x_2^{-1}(t)) = (\mathrm{pr}\, x_2)(x_2^{-1}(t)) = g x_2^{-1}(t).$$

Now, if we let $x = x_2 x_1|_U$, where $U = x_1^{-1}(U_2) \cap U_f$, we get a chart (x, U) such that

$$yfx^{-1}(t) = yfx_1^{-1}x_2^{-1}(t) = g x_2^{-1}(t) = \mathrm{pr}(t)$$

for all $t \in x(U) \subseteq U_2'$.

To end the proof, we need to cover the case when $y(q) = q' \neq 0$. Consider the translation $T\colon \mathbf{R}^n \cong \mathbf{R}^n$ given by $T(v) = v - q'$. Then (Ty, V) is a chart with $Ty(q) = 0$, so that the argument above gives us a chart (x, U) with $Tyfx^{-1}(s) = \mathrm{pr}(s)$, or equivalently, $yfx^{-1}(s) = \mathrm{pr}(s) + q'$. Letting $S\colon \mathbf{R}^m \cong \mathbf{R}^m$ be defined by $S(t) = t + (q', 0)$ we get that, when we exchange the chart (x, U) with (Sx, U), we have that

$$yf(Sx)^{-1}(t) = yfx^{-1}S^{-1}(t) = \mathrm{pr}(S^{-1}(t)) + q' = \mathrm{pr}(t)$$

for $t \in Sx(U)$, as desired. ■

Remember, if this is your first reading, you may now jump directly to Exercise 4.3.4 and come back to the full version of the rank theorem later (you will need it).

In the formulation of the rank theorem we give below, the two last cases are the extreme situations where the rank is maximal (and hence constant). Notice that the conditions vary quite subtly in terms of what charts can be arbitrary and what charts exist only as a consequence of the theorem.

In the first version, where the condition on the rank is weakest, we need to be able to tweak a given chart ever so slightly by means of a permutation. If $\sigma: \{1, \dots, n\} \to \{1, \dots, n\}$ is a bijection (a *permutation*), we refer to the diffeomorphism sending $(t_1, \dots, t_n)$ to $(t_{\sigma^{-1}(1)}, \dots, t_{\sigma^{-1}(n)})$ as a *permutation of the coordinates* corresponding to σ; which in an abuse of notation is also denoted $\sigma: \mathbf{R}^n \to \mathbf{R}^n$. When you start composing such permutations, you will appreciate that we insisted on the inverse of σ in the indices.

Exercise 4.3.2 The permutation of the coordinates is given by multiplication by a permutation matrix, and as a sanity check you might want to find out exactly what matrix this is.

Theorem 4.3.3 (The Rank Theorem) *Let M and N be smooth manifolds of dimensions $\dim(M) = m$ and $\dim(N) = n$, and let $\bar{f}: (M, p) \to (N, q)$ be a germ.*

1. *If $\bar{f}$ is of rank $\geq r$, then for any chart (z, V) for N with $q \in V$ there exists a chart (x, U) for M with $p \in U$ and permutation $\sigma: \mathbf{R}^n \to \mathbf{R}^n$ such that*

$$\overline{\mathrm{pr}\,\sigma\,zfx^{-1}} = \overline{\mathrm{pr}},$$

   where pr is the projection onto the first r coordinates: $\mathrm{pr}(t_1, \dots, t_m) = (t_1, \dots, t_r)$.
2. *If $\bar{f}$ has constant rank r, then there exist charts (x, U) for M and (y, V) for N with $p \in U$ and $q \in V$ such that*

$$\overline{yfx^{-1}} = \overline{i\,\mathrm{pr}},$$

   where $i\,\mathrm{pr}(t_1, \dots, t_m) = (t_1, \dots, t_r, 0, \dots, 0)$.
3. *If $\bar{f}$ is of rank n (and so $m \geq n$), then for any chart (y, V) for N with $f(p) \in V$, there exists a chart (x, U) for M with $p \in U$ such that*

$$\overline{yfx^{-1}} = \overline{\mathrm{pr}},$$

   where $\mathrm{pr}(t_1, \dots, t_m) = (t_1, \dots, t_n)$.

4. *If $\bar{f}$ is of rank m (and so m $\leq$ n), then for any chart (x, U) for M with $p \in U$ there exists a chart (y, V) for N with $f(p) \in V$ such that*

$$\overline{yfx^{-1}} = \bar{i},$$

where $i(t_1, \ldots, t_m) = (t_1, \ldots, t_m, 0, \ldots, 0)$.

Proof.   This is a local question: if we start with arbitrary charts, we will fix them up so that we have the theorem. Hence we may just as well assume that $(M, p) = (\mathbf{R}^m, 0)$ and $(N, f(p)) = (\mathbf{R}^n, 0)$, that f is a representative of the germ, and that the Jacobian $Df(0)$ has the form

$$Df(0) = \begin{bmatrix} A & B \\ C & D \end{bmatrix},$$

where A is an invertible $r \times r$ matrix. This is where we use that we may permute the coordinates: at the outset there was no guarantee that the upper left $r \times r$ matrix A was invertible: we could permute the columns by choosing x wisely (except in the fourth part, where x is fixed, but where this is unnecessary since $r = m$), but the best we could guarantee without introducing the σ was that there would be an invertible $r \times r$ matrix somewhere in the first r columns. For the third part of the theorem, this is unnecessary since $r = n$.

Let $f_i = \text{pr}_i f$, and for the first, second and third parts define $x \colon (\mathbf{R}^m, 0) \to (\mathbf{R}^m, 0)$ by

$$x(t) = (f_1(t), \ldots, f_r(t), t_{r+1}, \ldots, t_m)$$

(where $t_j = \text{pr}_j(t)$). Then

$$Dx(0) = \begin{bmatrix} A & B \\ 0 & I \end{bmatrix}$$

and so $\det Dx(0) = \det(A) \neq 0$. By the inverse function theorem, 4.2.2, $\bar{x}$ is an invertible germ with inverse $\bar{x}^{-1}$, and, as spelled out in Corollary 4.2.3, is represented by a chart. Choose a representative for $\bar{x}^{-1}$, which we, by a slight abuse of notation, will call x^{-1}. Since for sufficiently small $t \in M = \mathbf{R}^m$ we have

$$(f_1(t), \ldots, f_n(t)) = f(t) = fx^{-1}x(t) = fx^{-1}(f_1(t), \ldots, f_r(t), t_{r+1}, \ldots, t_m),$$

we see that

$$fx^{-1}(t) = (t_1, \ldots, t_r, f_{r+1}x^{-1}(t), \ldots, f_nx^{-1}(t))$$

and we have proven the first and third parts of the rank theorem.

For the second part, assume $\text{rk } Df(t) = r$ for all t. Since $\bar{x}$ is invertible

$$D(fx^{-1})(t) = Df(x^{-1}(t))D(x^{-1})(t)$$

also has rank r for all t in the domain of definition. Note that

$$D(fx^{-1})(t) = \begin{bmatrix} I & 0 \\ \cdots\cdots\cdots\cdots\cdots\cdots\cdots \\ [D_j(f_ix^{-1})(t)]_{\substack{i=r+1,\ldots,n \\ j=1,\ldots m}} \end{bmatrix},$$

so since the rank is exactly r we must have that the lower right-hand $(n-r)\times(m-r)$ matrix

$$\left[D_j(f_ix^{-1})(t) \right]_{\substack{r+1 \le i \le n \\ r+1 \le j \le m}}$$

is the zero matrix (which says that "for $i > r$, the function f_ix^{-1} does not depend on the last $m-r$ coordinates of the input"). Define $\bar{y}\colon (\mathbf{R}^n, 0) \to (\mathbf{R}^n, 0)$ by setting

$$y(t) = \left(t_1, \ldots, t_r, t_{r+1} - f_{r+1}x^{-1}(\bar{t}), \ldots, t_n - f_nx^{-1}(\bar{t}) \right),$$

where $\bar{t} = (t_1, \ldots, t_r, 0, \ldots, 0)$. Then

$$Dy(t) = \begin{bmatrix} I & 0 \\ ? & I \end{bmatrix},$$

so $\bar{y}$ is invertible and $\overline{yfx^{-1}}$ is represented by

$$t = (t_1, \ldots, t_m) \mapsto \left(t_1, \ldots, t_r, f_{r+1}x^{-1}(t) - f_{r+1}x^{-1}(\bar{t}), \ldots, f_nx^{-1}(t) - f_nx^{-1}(\bar{t}) \right)$$
$$= (t_1, \ldots, t_r, 0, \ldots, 0),$$

where the last equation holds since $D_j(f_ix^{-1})(t) = 0$ for $r < i \le n$ and $r < j \le m$ and so $f_ix^{-1}(t) - f_ix^{-1}(\bar{t}) = 0$ for $r < i \le n$ and t close to the origin.

For the fourth part, we need to shift the entire burden to the chart on $N = \mathbf{R}^n$. Consider the germ $\bar{\eta}\colon (\mathbf{R}^n, 0) \to (\mathbf{R}^n, 0)$ represented by $\eta(t) = (0, \ldots, 0, t_{m+1}, \ldots, t_n) + f(t_1, \ldots, t_m)$. Since

$$D\eta(0) = \begin{bmatrix} A & 0 \\ C & I \end{bmatrix}$$

is invertible, $\bar{\eta}$ is invertible. Let $\bar{y} = \bar{\eta}^{-1}$ and let y be the corresponding diffeomorphism. Since $\bar{f}$ is represented by $(t_1, \ldots, t_m) \mapsto \eta(t_1, \ldots, t_m, 0, \ldots, 0)$, we get that $\bar{y}\bar{f}$ is represented by $(t_1, \ldots, t_m) \mapsto (t_1, \ldots, t_m, 0, \ldots, 0)$, as required. ∎

Exercise 4.3.4

Let $f\colon M \to N$ be a smooth map between n-dimensional smooth manifolds. Assume that M is compact and that $q \in N$ a regular value. Prove that $f^{-1}(q)$ is a finite set and that there is an open neighborhood V around q such that for each $q' \in V$ we have that $f^{-1}(q') \cong f^{-1}(q)$.

Exercise 4.3.5 Prove the fundamental theorem of algebra: any non-constant complex polynomial P has a zero.

Exercise 4.3.6 Let $f: M \to M$ be smooth such that $f \circ f = f$ and M is connected. Prove that $f(M) \subseteq M$ is a submanifold. If you like point-set topology, prove that $f(M) \subseteq M$ is closed.

Note 4.3.7 It is a remarkable fact that any smooth manifold can be obtained by the method of Exercise 4.3.6 with M an open subset of Euclidean space and f some suitable smooth map. If you like algebra, then you might like to think that smooth manifolds are to open subsets of Euclidean spaces what projective modules are to free modules.

We will not be in a position to prove this, but the idea is as follows. Given a manifold T, choose a smooth imbedding $i: T \subseteq \mathbf{R}^N$ for some N (this is possible by virtue of the Whitney imbedding theorem, which we prove in Theorem 8.2.6 for T compact). Thicken $i(T)$ slightly to a "tubular neighborhood" U, which is an open subset of $\mathbf{R}^N$ together with a lot of structure (it is isomorphic to what we will later refer to as the "total space of the normal bundle" of the inclusion $i(T) \subseteq \mathbf{R}^N$), and in particular comes equipped with a smooth map $f: U \to U$ (namely the "projection $U \to i(T)$ of the bundle" composed with the "zero section $i(T) \to U$" – you'll recognize these words once we have talked about vector bundles) such that $f \circ f = f$ and $f(U) = i(T)$.

4.4 Regular Values

Since by Lemma 4.1.2 the rank cannot decrease locally, there are certain situations where constant rank is guaranteed, namely when the rank is maximal.

Definition 4.4.1 A smooth map $f: M \to N$ is

a *submersion* if rk $T_p f = \dim N$ (that is, $T_p f$ is surjective)
an *immersion* if rk $T_p f = \dim M$ ($T_p f$ is injective)

for all $p \in M$.

In these situations the third and/or fourth version in the Rank Theorem, 4.3.3, applies.

Note 4.4.2 To say that a map $f: M \to N$ is a submersion is equivalent to claiming that all points $p \in M$ are regular ($T_p f$ is surjective), which again is equivalent to claiming that all $q \in N$ are regular values (values that are not hit are regular by definition).

Theorem 4.4.3 *Let*

$$f : M \to N$$

be a smooth map where M is $(n + k)$-dimensional and N is n-dimensional. If $q \in N$ is a regular value and $f^{-1}(q)$ is not empty, then

$$f^{-1}(q) \subseteq M$$

is a k-dimensional smooth submanifold.

Proof. Let $p \in f^{-1}(q)$. We must display a chart (x, W) around p such that

$$x(W \cap f^{-1}(q)) = x(W) \cap (\mathbf{R}^k \times \{0\}).$$

Since p is regular, the rank theorem, 4.3.3(3) (aka Lemma 4.3.1), implies that there are charts (x, U) and (y, V) around p and q such that $x(p) = 0$, $y(q) = 0$ and

$$yfx^{-1}(t_1, \ldots, t_{n+k}) = (t_1, \ldots, t_n), \text{ for } t \in x(U \cap f^{-1}(V))$$

(moving $x(p)$ and $y(q)$ to the origins does not mess up the conclusion). Let $W = U \cap f^{-1}(V)$, and note that $W \cap f^{-1}(q) = (yf|_W)^{-1}(0)$. Then

$$\begin{aligned} x(W \cap f^{-1}(q)) &= \left(yfx^{-1}|_{x(W)}\right)^{-1}(0) \\ &= \{(0, \ldots, 0, t_{n+1}, \ldots, t_{n+k}) \in x(W)\} \\ &= x(W) \cap (\{0\} \times \mathbf{R}^k) \end{aligned}$$

and so (permuting the coordinates) $f^{-1}(q) \subseteq M$ is a k-dimensional submanifold as claimed. ∎

Exercise 4.4.4 Give a new proof which shows that $S^n \subset \mathbf{R}^{n+1}$ is a smooth submanifold.

Note 4.4.5 Not all submanifolds can be realized as the inverse image of a regular value of some map (for instance, the central circle in the Möbius band is not the inverse image of a regular value of any function, c.f. Example 5.1.4), but the theorem still gives a rich source of important examples of submanifolds.

Example 4.4.6 Consider the *special linear group*

$$\mathrm{SL}_n(\mathbf{R}) = \{A \in \mathrm{GL}_n(\mathbf{R}) \mid \det(A) = 1\}.$$

We show that $\mathrm{SL}_2(\mathbf{R})$ is a three-dimensional manifold. The determinant function is given by

$$\det : M_2(\mathbf{R}) \to \mathbf{R},$$

$$A = \begin{bmatrix} a_{11} & a_{12} \\ a_{21} & a_{22} \end{bmatrix} \mapsto \det(A) = a_{11}a_{22} - a_{12}a_{21}$$

and so, with the obvious coordinates $M_2(\mathbf{R}) \cong \mathbf{R}^4$ (sending A to $[a_{11}\, a_{12}\, a_{21}\, a_{22}]^{\mathrm{T}}$), we have that

$$D(\det)(A) = \begin{bmatrix} a_{22} & -a_{21} & -a_{12} & a_{11} \end{bmatrix}.$$

Hence the determinant function has rank 1 at all matrices, except the zero matrix, and in particular 1 is a regular value.

Exercise 4.4.7   Show that $\mathrm{SL}_2(\mathbf{R})$ is diffeomorphic to $S^1 \times \mathbf{R}^2$.

Exercise 4.4.8   If you have the energy, you may prove that $\mathrm{SL}_n(\mathbf{R})$ is an $(n^2 - 1)$-dimensional manifold.

Example 4.4.9   The subgroup $\mathrm{O}(n) \subseteq \mathrm{GL}_n(\mathbf{R})$ of *orthogonal matrices* is a submanifold of dimension $n(n-1)/2$.

To see this, recall that $A \in M_n(\mathbf{R})$ is orthogonal if and only if $A^{\mathrm{T}}A = I$. Note that $A^{\mathrm{T}}A$ is always symmetric. The space $\mathrm{Sym}(n)$ of all symmetric matrices is diffeomorphic to $\mathbf{R}^{n(n+1)/2}$ (the entries on and above the diagonal can be chosen arbitrarily, and will then determine the remaining entries uniquely). We define a map

$$f \colon M_n(\mathbf{R}) \to \mathrm{Sym}(n),$$
$$A \mapsto A^{\mathrm{T}}A$$

which is smooth (since matrix multiplication and transposition are smooth), and such that

$$\mathrm{O}(n) = f^{-1}(I).$$

We must show that I is a regular value, and we offer two proofs, one computational using the Jacobi matrix, and one showing more directly that $T_A f$ is surjective for all $A \in \mathrm{O}(n)$. We present both proofs; the first one since it is very concrete, and the second one since it is short and easy to follow.

First we give the Jacobian argument. We use the usual chart $M_n(\mathbf{R}) \cong \mathbf{R}^{n^2}$ by listing the entries in lexicographical order, and the chart

$$\mathrm{pr} \colon \mathrm{Sym}(n) \cong \mathbf{R}^{n(n+1)/2}$$

with $\mathrm{pr}_{ij}[A] = a_{ij}$ if $A = [a_{ij}]$ (also in lexicographical order) defined only for $1 \leq i \leq j \leq n$. Then $\mathrm{pr}_{ij} f([A]) = \sum_{k=1}^{n} a_{ki} a_{kj}$, and a straightforward calculation yields that

$$D_{kl}\, \mathrm{pr}_{ij} f(A) = \begin{cases} a_{ki} & i < j = l \\ a_{kj} & l = i < j \\ 2a_{kl} & i = j = l \\ 0 & \text{otherwise.} \end{cases}$$

In particular

$$D_{kl} \, \mathrm{pr}_{ij} f(I) = \begin{cases} 1 & k = i < j = l \\ 1 & l = i < j = k \\ 2 & i = j = k = l \\ 0 & \text{otherwise} \end{cases}$$

and rk $Df(I) = n(n+1)/2$ since $Df(I)$ is on echelon form, with no vanishing rows. (As an example, for $n = 2$ and $n = 3$ the Jacobi matrices are

$$\begin{bmatrix} 2 & & \\ & 1 & 1 \\ & & 2 \end{bmatrix} \text{ and } \begin{bmatrix} 2 & & & & & \\ & 1 & & 1 & & \\ & & 1 & & & 1 \\ & & & 2 & & \\ & & & & 1 & 1 \\ & & & & & 2 \end{bmatrix}$$

(in the first matrix the columns are the partial derivatives in the 11, 12, 21 and 22 variables, and the rows are the projection on the 11 12 and 22 factors; likewise in the second one).)

For any $A \in \mathrm{GL}_n(\mathbf{R})$ we define the diffeomorphism $L_A \colon M_n(\mathbf{R}) \to M_n(\mathbf{R})$ by $L_A(B) = A \cdot B$. Note that if $A \in O(n)$ then

$$f(L_A(B)) = f(AB) = (AB)^{\mathrm{T}} AB = B^{\mathrm{T}} A^{\mathrm{T}} AB = B^{\mathrm{T}} B = f(B),$$

and so, by the chain rule and the fact that $D(L_A)(B) \cdot = A \cdot$, we get that

$$Df(I) \cdot = D(fL_A)(I) \cdot = D(f)(L_A I) \cdot D(L_A)(I) \cdot = D(f)(A) \cdot A \cdot,$$

implying that rk $D(f)(A) = n(n+1)/2$ for all $A \in O(n)$. This means that A is a regular point for all $A \in O(n) = f^{-1}(I)$, and so I is a regular value, and $O(n)$ is a submanifold of dimension

$$n^2 - n(n+1)/2 = n(n-1)/2.$$

For the other proof of the fact that I is a regular value, we use that $M_n(\mathbf{R})$ and $\mathrm{Sym}(n)$ are Euclidean spaces. In particular, any tangent vector in $T_A M_n(\mathbf{R})$ is represented by a linear curve

$$v_B(s) = A + sB, \qquad B \in M_n(\mathbf{R}), \quad s \in \mathbf{R}.$$

We have that

$$f v_B(s) = (A + sB)^{\mathrm{T}}(A + sB) = A^{\mathrm{T}} A + s(A^{\mathrm{T}} B + B^{\mathrm{T}} A) + s^2 B^{\mathrm{T}} B$$

and so

$$T_A f[v_B] = [f v_B] = [\gamma_B],$$

where $\gamma_B(s) = A^{\mathrm{T}} A + s(A^{\mathrm{T}} B + B^{\mathrm{T}} A)$. Similarly, any tangent vector in $T_I \mathrm{Sym}(n)$ is in the equivalence class of a linear curve

$$\alpha_C(s) = I + sC$$

for C a symmetric matrix. If A is orthogonal, we see that $\gamma_{\frac{1}{2}AC} = \alpha_C$, and so $T_A f[v_{\frac{1}{2}AC}] = [\alpha_C]$, and $T_A f$ is surjective. Since this is true for any $A \in \mathrm{O}(n)$, we get that I is a regular value.

Note 4.4.10 The multiplication

$$\mathrm{O}(n) \times \mathrm{O}(n) \to \mathrm{O}(n)$$

is smooth (since multiplication of matrices is smooth in $M_n(\mathbf{R}) \cong \mathbf{R}^{n^2}$, and we have the result of Exercise 2.5.16), so $\mathrm{O}(n)$ is a Lie group. The same of course applies to $\mathrm{SL}_n(\mathbf{R})$.

Exercise 4.4.11 Prove that the *unitary group*

$$\mathrm{U}(n) = \{A \in \mathrm{GL}_n(\mathbf{C}) \mid \bar{A}^{\mathrm{T}} A = I\}$$

is a Lie group of dimension n^2.

Exercise 4.4.12 Prove that $\mathrm{O}(n)$ is compact and has two connected components. The component consisting of matrices of determinant 1 is called $\mathrm{SO}(n)$, the *special orthogonal group*.

Exercise 4.4.13 Prove that $\mathrm{SO}(2)$ is diffeomorphic to S^1, and that $\mathrm{SO}(3)$ is diffeomorphic to the real projective 3-space.

Note 4.4.14 It is a beautiful fact that, if G is a Lie group (e.g., $\mathrm{GL}_n(\mathbf{R})$) and $H \subseteq G$ is a closed subgroup (i.e., a closed subset which is closed under multiplication and such that if $h \in H$ then $h^{-1} \in H$), then $H \subseteq G$ is a "Lie subgroup". We will not prove this fact, (see, e.g., Theorem 10.15 of [20]), but note that it implies that **all** closed matrix groups such as $\mathrm{O}(n)$ are Lie groups since $\mathrm{GL}_n(\mathbf{R})$ is.

Example 4.4.15 Consider the map $f : S^1 \times S^1 \times S^1 \to \mathrm{SO}(3)$ uniquely defined by the composite $g : \mathbf{R}^3 \to S^1 \times S^1 \times S^1 \to \mathrm{SO}(3) \subseteq M_3(\mathbf{R})$ sending (α, β, γ) to

$$\begin{bmatrix} \cos\gamma & -\sin\gamma & 0 \\ \sin\gamma & \cos\gamma & 0 \\ 0 & 0 & 1 \end{bmatrix} \begin{bmatrix} 1 & 0 & 0 \\ 0 & \cos\beta & -\sin\beta \\ 0 & \sin\beta & \cos\beta \end{bmatrix} \begin{bmatrix} \cos\alpha & -\sin\alpha & 0 \\ \sin\alpha & \cos\alpha & 0 \\ 0 & 0 & 1 \end{bmatrix}.$$

A quick calculation shows that the rank of this map is 3, unless $\sin\beta = 0$, in which case the rank is 2. (Do this calculation!) Hence all points on $S^1 \times S^1 \times S^1$ are regular except those in the two sub-tori $S^1 \times \{\pm 1\} \times S^1 \subseteq S^1 \times S^1 \times S^1$ with middle coordinate 1 or -1 ($\beta = 0$ or $\beta = \pi$). (Why? Explain why the rank of the composite g gives the rank of f.) For instance, on the sub-torus with $\beta = 0$, the rotation is simply $\alpha + \gamma$ around the z-axis.

Hence, around any point away from these two tori, f is a local diffeomorphism, and can be used as "coordinates" for $\mathrm{SO}(3)$, and the angles of the set (α, β, γ)

(with $\beta \in (0, \pi)$ – ensuring uniqueness) are called the *Euler angles*, representing this rotation. For a fuller treatment of Euler angles, check Wikipedia.

Euler angles are used, e.g., in computer graphics and in flight control to represent rotations. However, a rotation in the image of the critical sub-tori has an inverse image consisting of an entire circle (for instance, $f^{-1}(f(z, 1, w)) = \{(zy^{-1}, 1, yw) \mid y \in S^1\}$), a situation which is often referred to as the *gimbal lock* and is considered highly undesirable. This name derives from navigation, where one uses a device called an inertial measurement unit (IMU) to keep a reference frame to steer by (it consists of three gimbals mounted inside each other at right angles to provide free rotation in all directions with gyroscopes in the middle to provide inertia fixing the reference frame). The map f above gives the correspondence between the rotation in question and the angles in the gimbals. However, at the critical value of f – the gimbal lock – the IMU fails to work, causing a loss of reference frame. Hence a plane has to avoid maneuvering too close to the gimbal lock.

See www.hq.nasa.gov/alsj/gimbals.html giving some background on the worries the gimbal lock caused NASA's Apollo mission.

Exercise 4.4.16

A *k-frame* in $\mathbf{R}^n$ is a k-tuple of orthonormal vectors in $\mathbf{R}^n$. Define the *Stiefel manifold* V_n^k (named after Eduard Stiefel (1909–1978))[1] as the subset

$$V_n^k = \{k\text{-frames in } \mathbf{R}^n\}$$

of $\mathbf{R}^{nk}$. Show that V_n^k is a compact smooth $(nk - k(k+1)/2)$-dimensional manifold. Note that V_n^1 may be identified with S^{n-1}.

Note 4.4.17 In the literature you will often find a different definition, where a k-frame is just a k-tuple of linearly independent vectors. Then the Stiefel manifold is an open subset of $M_{n \times k}(\mathbf{R})$, and so is clearly a smooth manifold – but this time of dimension nk.

A k-frame defines a k-dimensional linear subspace of $\mathbf{R}^n$. The Grassmann manifold $\mathrm{Gr}(k, \mathbf{R}^n)$ of Example 2.3.15 has as underlying set the set of k-dimensional linear subspaces of $\mathbf{R}^n$, and we get a quotient map $V_n^k \to \mathrm{Gr}(k, \mathbf{R}^n)$. In particular, the quotient map $V_{n+1}^1 \to \mathrm{Gr}(1, \mathbf{R}^{n+1})$ may be identified with the projection $S^n \to \mathbf{RP}^n$.

Exercise 4.4.18

Let P_n be the space of degree-n polynomials. Show that the space of solutions in P_3 of the equation

$$(y'')^2 - y' + y(0) + xy'(0) = 0$$

is a one-dimensional submanifold of P_3.

....

[1] https://en.wikipedia.org/wiki/Eduard_Stiefel

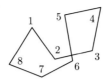

Figure 4.4. A labeled flexible 8-gon in $\mathbf{R}^2$.

Exercise 4.4.19 Formulate a more interesting exercise along the lines of the previous one, and solve it.

Exercise 4.4.20 Let $A \in M_n(\mathbf{R})$ be a symmetric matrix. For what values of $a \in \mathbf{R}$ is the *quadric*

$$M_a^A = \{p \in \mathbf{R}^n \mid p^{\mathrm{T}} A p = a\}$$

an $(n-1)$-dimensional smooth manifold?

Exercise 4.4.21 In a chemistry book I found the van der Waals equation, which gives a relationship between the temperature T, the pressure p and the volume V, which supposedly is somewhat more accurate than the ideal gas law $pV = nRT$ (n is the number of moles of gas, R is a constant). Given the relevant positive constants a and b, prove that the set of points $(p, V, T) \in (0, \infty) \times (nb, \infty) \times (0, \infty)$ satisfying the equation

$$\left(p - \frac{n^2 a}{V^2}\right)(V - nb) = nRT$$

is a smooth submanifold of $\mathbf{R}^3$.

Exercise 4.4.22 Consider the set $F_{n,k}$ of *labeled flexible n-gons* in $\mathbf{R}^k$. A labeled flexible n-gon (e.g., the one shown in Figure 4.4) is what you get if you join $n > 2$ straight lines of unit length to a closed curve and label the vertices from 1 to n.

Let n be odd and $k = 2$. Show that $F_{n,2}$ is a smooth submanifold of $\mathbf{R}^2 \times (S^1)^{n-1}$ of dimension n.

Exercise 4.4.23 For odd n, prove that the set of *non-self-intersecting* flexible n-gons in $\mathbf{R}^2$ is a manifold.

4.5 Transversality

In Theorem 4.4.3 we learned about regular values, and inverse images of these. Often interesting submanifolds naturally occur not as inverse images of points, but as inverse images of submanifolds. A spectacular example appears in Note 4.5.8, where the non-diffeomorphic smooth manifolds homeomorphic to S^7 are constructed in this way. How is one to guarantee that the inverse image of a submanifold is a submanifold? The relevant term is *transversality*.

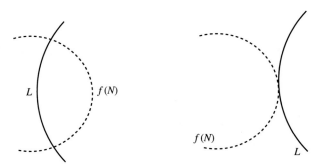

Figure 4.5. The picture to the left is a typical transverse situation, whereas in the picture to the right f definitely can't be transverse to L since $\mathrm{Im}\{T_p f\}$ and $T_{f(p)}L$ span only a one-dimensional space. Beware that pictures like this can be misleading, since the situation to the left fails to be transverse if f slows down at a point p of intersection, i.e., if $T_p f = 0$.

Definition 4.5.1 Let $f\colon N \to M$ be a smooth map and $L \subset M$ a smooth submanifold. We say that f is *transverse* to $L \subset M$ if for all $p \in f^{-1}(L)$ the image of $T_p f$ and $T_{f(p)}L$ together span $T_{f(p)}M$. See Figure 4.5.

Example 4.5.2 An important example is given by the case when f is the inclusion of a submanifold: we say that two smooth submanifolds N and L of M are *transverse* if for all $p \in N \cap L$ the subspaces $T_p N$ and $T_p L$ together span all of $T_p M$ (note that if $f\colon N \subseteq M$ is the inclusion, then $f^{-1}(L) = N \cap L$).

At the other extreme, we may consider two smooth maps $f\colon N \to M$, $g\colon L \to M$, see Exercise 4.7.11.

Exercise 4.5.3 Let $z = (a, b) \in S^1 \subseteq \mathbf{R}^2$ and let $N_z = \{(a, y)\,|\,y \in \mathbf{R}\}$ be the vertical line intersecting the circle at z. When is $S^1 \subseteq \mathbf{R}^2$ transverse to $N_z \subseteq \mathbf{R}^2$?

Note 4.5.4 If $L = \{q\}$ in the definition above, we recover the definition of a regular point.

Another common way of expressing transversality is to say that for all $p \in f^{-1}(L)$ the induced map

$$T_p N \xrightarrow{\ T_p f\ } T_{f(p)}M \longrightarrow T_{f(p)}M/T_{f(p)}L$$

is surjective. Here $T_{f(p)}M/T_{f(p)}L$ is the quotient space: recall that, if W is a subspace of a vector space V, then the quotient space V/W is the vector space you get from V by dividing out by the equivalence relation $v \sim v + w$ for $v \in V$ and $w \in W$. The vector space structure on V/W is defined by demanding that the map $V \to V/W$ sending a vector to its equivalence class is linear.

Note that the definition of transversality refers only to points in $f^{-1}(L)$, so if $f(N) \cap L = \emptyset$ the condition is vacuous and f and L are transverse, as shown in Figure 4.6.

Furthermore, if f is a submersion (i.e., $T_p f$ is always surjective), then f is transverse to all submanifolds.

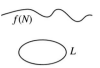

Figure 4.6. A map is always transverse to a submanifold its image does not intersect.

Theorem 4.5.5 *Assume that the smooth map $f: N \to M$ is transverse to a k-codimensional smooth submanifold $L \subseteq M$ and that $f(N) \cap L \neq \emptyset$. Then $f^{-1}(L) \subseteq N$ is a k-codimensional smooth submanifold.*

Proof. Let $q \in L$ and $p \in f^{-1}(q)$. Choose a chart (y, V) around q such that $y(q) = 0$ and such that

$$y(L \cap V) = y(V) \cap (\mathbf{R}^{m-k} \times \{0\}),$$

where m is the dimension of M. Let $\pi: \mathbf{R}^m \to \mathbf{R}^k$ be the projection $\pi(t_1, \ldots, t_m) = (t_{m-k+1}, \ldots, t_m)$. Consider the commutative diagram

$$
\begin{array}{ccccccc}
T_p N & \xrightarrow{T_p f} & T_q M & \xrightarrow{T_q y} & T_0 \mathbf{R}^m & \xrightarrow{T_0 \pi} & T_0 \mathbf{R}^k, \\
 & \searrow & \downarrow{\scriptstyle\text{projection}} & & \downarrow{\scriptstyle\text{projection}} & \nearrow & \\
 & {\scriptstyle\text{surjective}} & T_q M / T_q L & \xrightarrow[\cong]{} & T_0 \mathbf{R}^m / T_0 \mathbf{R}^{m-k} & &
\end{array}
$$

where the left diagonal map is surjective by the transversality assumption and the right diagonal map is the isomorphism making the triangle commute. Then we get that p is a regular point of the composite

$$U = f^{-1}(V) \xrightarrow{f|_U} V \xrightarrow{y} y(V) \xrightarrow{\pi|_{y(V)}} \mathbf{R}^k.$$

This is true for any $p \in f^{-1}(V \cap L)$ so $0 \in \mathbf{R}^k$ is a regular value. Hence

$$(\pi y f|_U)^{-1}(0) = f^{-1} y^{-1} \pi^{-1}(0) \cap U = f^{-1}(L) \cap U$$

is a submanifold of codimension k in U. Varying q, we therefore get that $f^{-1}(L) \subseteq N$ is a k-codimensional submanifold. $\blacksquare$

Corollary 4.5.6 *If N and L are transverse submanifolds of M, then $N \cap L$ is a smooth $(\dim N + \dim L - \dim M)$-dimensional smooth submanifold of N.*

Exercise 4.5.7 Let $n \geq 3$ and let $a_0, \ldots, a_n$ be integers greater than 1. Consider $f: \mathbf{C}^{n+1} \setminus \{0\} \to \mathbf{C}$ given by $f(z_0, \ldots, z_n) = \sum_k z_k^{a_k}$. Then $L = f^{-1}(0)$ is a $2n$-dimensional submanifold of $\mathbf{C}^{n+1}$ which is transverse to the sphere $S^{2n+1} \subseteq \mathbf{C}^{n+1}$.

Note 4.5.8 The intersection $W^{2n-1}(a_0, \ldots, a_n) = L \cap S^{2n+1}$ of Exercise 4.5.7 is called the *Brieskorn manifold*. Brieskorn showed [3] that

$W^{4m-1}(3, 6r - 1, 2, \ldots, 2)$ is homeomorphic to S^{4m-1}. However if $m = 2$ (so that $4m - 1 = 7$), then $W^7(3, 6r - 1, 2, 2, 2)$ for $r = 1, \ldots, 28$ is a complete list of the 28 smooth oriented structures on the seven-dimensional sphere mentioned in Note 2.3.7! The situation for higher dimensions varies according to whether we are in dimension $+1$ or -1 modulo 4.

4.6 Sard's Theorem[2]

As commented earlier, the regular points are dense. Although this is good to know and important for many applications, we will not need this fact, and are content to cite the precise statement and let the proof be a guided sequence of exercises. Proofs can be found in many references, for instance in Chapter 3 of [15].

Theorem 4.6.1 (Sard) *Let $f : M \rightarrow N$ be a smooth map. The set of critical values has measure zero.*

Recall that a subset $C \subseteq \mathbf{R}^n$ has *measure zero* if for every $\epsilon > 0$ there is a sequence of closed cubes $\{C_i\}_{i \in \mathbf{N}}$ with $C \subseteq \bigcup_{i \in \mathbf{N}} C_i$ and $\sum_{i \in \mathbf{N}}$ volume$(C_i) < \epsilon$.

In this definition it makes no essential difference if one uses open or closed cubes, rectangles or balls instead of closed cubes.

Exercise 4.6.2 Any open subset U of $\mathbf{R}^n$ is a countable union of closed balls.

Exercise 4.6.3 Prove that a countable union of measure-zero subsets of $\mathbf{R}^n$ has measure zero.

Exercise 4.6.4 Let $f : U \rightarrow \mathbf{R}^m$ be a smooth map, where $U \subseteq \mathbf{R}^m$ is an open subset. Prove that, if $C \subseteq U$ has measure zero, then so does the image $f(C)$. Conclude that a diffeomorphism $f : U \rightarrow U'$ between open subsets of Euclidean spaces provides a one-to-one correspondence between the subsets of measure zero in U and in U'.

Definition 4.6.5 Let $(M, \mathcal{A})$ be a smooth n-dimensional manifold and $C \subseteq M$ a subset. We say that C has *measure zero* if for each $(x, U) \in \mathcal{A}$ the subset $x(C \cap U) \subseteq \mathbf{R}^n$ has measure zero.

Given a subatlas $\mathcal{B} \subseteq \mathcal{A}$, we see that by Exercise 4.6.4 it suffices to check that $x(C \cap U) \subseteq \mathbf{R}^n$ has measure zero for all $(x, U) \in \mathcal{B}$.

Exercise 4.6.6 An open cover of the closed interval $[0, 1]$ by subintervals contains a finite open subcover whose sum of diameters is less than or equal to 2.

----

[2] This material is not used in an essential way in the rest of the book. It is included for completeness, and for comparison with other sources.

Exercise 4.6.7 Prove Fubini's theorem. Let $C \subseteq \mathbf{R}^n$ be a countable union of compact subsets. Assume that for each $t \in \mathbf{R}$ the set

$$\{(t_1, \ldots, t_{n-1}) \in \mathbf{R}^{n-1} \mid (t_1, \ldots, t_{n-1}, t) \in C\} \subseteq \mathbf{R}^{n-1}$$

has measure zero. Then C has measure zero.

Exercise 4.6.8 Show that Sard's theorem follows if you show the following statement. Let $f : U \to \mathbf{R}^n$ be smooth where $U \subseteq \mathbf{R}^m$ is open, and let C be the set of critical points. Then $f(C) \subseteq \mathbf{R}^n$ has measure zero.

Let $f : U \to \mathbf{R}^n$ be smooth where $U \subseteq \mathbf{R}^m$ is open, and let C be the set of critical points. For $i > 0$, let C_i be the set of points $p \in U$ such that all partial derivatives of order less than or equal to i vanish, and let $C_0 = C$.

Exercise 4.6.9 Assume Sard's theorem is proven for manifolds of dimension less than m. Prove that $f(C_0 - C_1)$ has measure zero.

Exercise 4.6.10 Assume Sard's theorem is proven for manifolds of dimension less than m. Prove that $f(C_i - C_{i+1})$ has measure zero for all $i > 0$.

Exercise 4.6.11 Assume Sard's theorem is proven for manifolds of dimension less than m. Prove that $f(C_k)$ has measure zero for $nk \geq m$.

Exercise 4.6.12 Prove Sard's theorem.

4.7 Immersions and Imbeddings

We are finally closing in on the promised effective definition of submanifolds, or rather, of imbeddings. The condition of being an immersion is a readily checked property, since we merely have to check the derivatives at every point. The rank theorem states that in some sense "locally" immersions are imbeddings. But how much more do we need? Obviously, an imbedding is injective.

Something more is needed, as we see from the following example.

Example 4.7.1 Consider the injective smooth map

$$f : (0, 3\pi/4) \to \mathbf{R}^2$$

given by

$$f(t) = \sin(2t)(\cos t, \sin t).$$

Then

$$Df(t) = 2[(1 - 3\sin^2 t)\cos t, (3\cos^2 t - 1)\sin t]$$

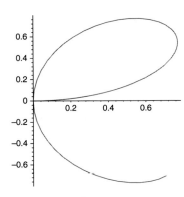

Figure 4.7. The image of f is a subspace of $\mathbf{R}^2$.

is never zero and f is an immersion.

However, $(0, 3\pi/4) \to \mathrm{Im}\{f\}$ is not a homeomorphism, where

$$\mathrm{Im}\{f\} = f((0, 3\pi/4)) \subseteq \mathbf{R}^2$$

has the subspace topology (Figure 4.7). For, if it were a homeomorphism, then

$$f((\pi/4, 3\pi/4)) \subseteq \mathrm{Im}\{f\}$$

would be open (for the inverse to be continuous). But any open ball around $(0, 0) = f(\pi/2)$ in $\mathbf{R}^2$ must contain a piece of $f((0, \pi/4))$, so $f((\pi/4, 3\pi/4)) \subseteq \mathrm{Im}\{f\}$ is not open.

Hence f is not an imbedding.

Exercise 4.7.2 Let

$$\mathbf{R} \coprod \mathbf{R} \to \mathbf{R}^2$$

be defined by sending x in the first summand to $(x, 0)$ and y in the second summand to $(0, e^y)$. This is an injective immersion, but not an imbedding.

Exercise 4.7.3 Let

$$\mathbf{R} \coprod S^1 \to \mathbf{C}$$

be defined by sending x in the first summand to $(1 + e^x)e^{ix}$ and being the inclusion $S^1 \subseteq \mathbf{C}$ on the second summand. This is an injective immersion, but not an imbedding (Figure 4.8).

But, strangely enough, these examples exhibit the only thing that can go wrong: if an injective immersion is to be an imbedding, the map to the image has got to be a homeomorphism.

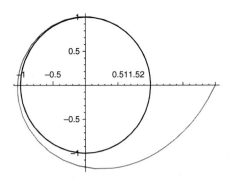

Figure 4.8. The image is not a submanifold of **C**.

Theorem 4.7.4 *If $f: M \to N$ is an immersion such that the induced map*

$$M \to \mathrm{Im}\{f\}$$

is a homeomorphism where $\mathrm{Im}\{f\} = f(M) \subseteq N$ has the subspace topology, then f is an imbedding.

Proof. Let $p \in M$. The rank theorem, 4.3.3(4), implies that there are charts

$$x_1: U_1 \to U_1' \subseteq \mathbf{R}^m$$

with $p \in U_1$ and $x_1(p) = 0$ and

$$y_1: V_1 \to V_1' \subseteq \mathbf{R}^{m+k}$$

with $f(p) \in V_1$ and $y_1(f(p)) = 0$, such that

$$y_1 f x_1^{-1}(t) = (t, 0) \in \mathbf{R}^m \times \mathbf{R}^k = \mathbf{R}^{m+k}$$

for all $t \in x_1(A)$, where $A = U_1 \cap f^{-1}(V_1)$, so that $x_1(A) = U_1' \cap x_1 f^{-1}(V_1)$.

Since V_1' is open, it contains open rectangles around the origin. Choose one such rectangle $V_2' = U' \times B \subseteq V_1'$ so that $U' \subseteq x_1(A)$ (see Figure 4.9).

Let $U = x_1^{-1}(U')$, $x = x_1|_U$ and $V_2 = y_1^{-1}(V_2')$.

Since $M \to f(M)$ is a homeomorphism, $f(U)$ is an open subset of $f(M)$, and since $f(M)$ has the subspace topology, $f(U) = W \cap f(M)$ where W is an open subset of N (here is the crucial point where complications as in Example 4.7.1 are excluded: there are no other "branches" of $f(M)$ showing up in W).

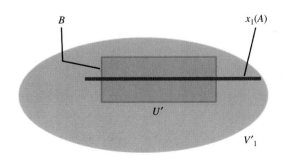

Figure 4.9.

Let $V = V_2 \cap W$, $V' = V_2' \cap y_1(W)$ and $y = y_1|_V$.

Then we see that $f(M) \subseteq N$ is a submanifold ($y(f(M) \cap V) = y(f(M) \cap W \cap V_2) = yf(U) = (\mathbf{R}^m \times \{0\}) \cap V'$), and that $M \to f(M)$ is a bijective local diffeomorphism (the constructed charts show that both $M \to f(M)$ and its inverse $f^{-1}|_{f(M)} \colon f(M) \to M$ are smooth around every point), and hence a diffeomorphism. ∎

We note the following useful corollary.

Corollary 4.7.5 *Let $f \colon M \to N$ be an injective immersion from a compact manifold M. Then f is an imbedding.*

Proof. We need only show that the continuous map $M \to f(M)$ is a homeomorphism. It is injective since f is, and clearly surjective. But from point-set topology (Theorem A.7.8) we know that it must be a homeomorphism since M is compact and $f(M)$ is Hausdorff ($f(M)$ is Hausdorff since it is a subspace of the Hausdorff space N). ∎

Those readers who struggled to get Exercise 2.5.17 right using only the definitions will appreciate the fact that Theorem 4.7.4 makes everything so much simpler:

Exercise 4.7.6 Show that the composite of two imbeddings is an imbedding.

Exercise 4.7.7 Let $a, b \in \mathbf{R}$, and consider the map

$$f_{a,b} \colon \mathbf{R} \to S^1 \times S^1,$$
$$t \mapsto (e^{iat}, e^{ibt}).$$

Show that $f_{a,b}$ is an immersion if either a or b is different from zero. Show that $f_{a,b}$ factors through an imbedding $S^1 \to S^1 \times S^1$ if and only if either $b = 0$ or a/b is rational (Figure 4.10).

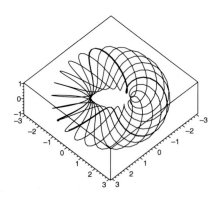

Figure 4.10. Part of the picture if $a/b = \pi$ (this goes on forever).

Exercise 4.7.8 Consider smooth maps

$$M \xrightarrow{\ i\ } N \xrightarrow{\ j\ } L.$$

Show that, if the composite ji is an imbedding, then i is an imbedding.

Example 4.7.9 As a last example of Corollary 4.7.5 we can redo Exercise 2.5.13 and see that

$$f: \mathbf{RP}^n \to \mathbf{RP}^{n+1},$$
$$[p] = [p_0, \ldots, p_n] \mapsto [p, 0] = [p_0, \ldots, p_n, 0]$$

is an imbedding: $\mathbf{RP}^n$ is compact, and f is injective and an immersion (check immersion in the standard charts or do it on spheres instead since the projection is a local diffeomorphism).

Exercise 4.7.10 Let M be a smooth manifold, and consider M as a subset by imbedding it as the diagonal in $M \times M$, i.e., as the set $\{(p, p) \in M \times M\}$: show that it is a smooth submanifold.

Exercise 4.7.11 Consider two smooth maps

$$M \xrightarrow{\ f\ } N \xleftarrow{\ g\ } L$$

Define the *fiber product*

$$M \times_N L = \{(p, q) \in M \times L \mid f(p) = g(q)\}$$

(topologized as a subspace of the product $M \times L$: notice that, if f and g are inclusions of subspaces, then $M \times_N L = M \cap L$). Assume that for all $(p, q) \in M \times_N L$ the subspaces spanned by the images of $T_p M$ and $T_q L$ equal all of $T_{f(p)} N$. Show that the fiber product $M \times_N L \subseteq M \times L$ is a smooth submanifold (of codimension equal to the dimension of N) such that the projections $M \times_N L \to M$ and $M \times_N L \to L$ are smooth.

Exercise 4.7.12 Let $\pi: E \to M$ be a submersion and $f: N \to M$ smooth. Let $E \times_M N$ be the fiber product of Exercise 4.7.11. Show that the projection $E \times_M N \to N$ is a submersion.

5 Vector Bundles

In this chapter we are going to collect all the tangent spaces of a manifold into a single object, the so-called *tangent bundle*.

5.0.1 The Idea

We defined the tangent space at a point in a smooth manifold by considering curves passing through the point (Figure 5.1). In physical terms, the tangent vectors are the velocity vectors of particles passing through our given point. But the particle will have velocities and positions at other times than the one at which it passes through our given point, and the position and velocity may depend continuously upon the time. Such a broader view demands that we are able to keep track of the points on the manifold and their tangent space, and understand how they change from point to point.

As a *set* the tangent bundle ought to be given by pairs (p, v), where $p \in M$ and $v \in T_p M$, i.e.,

$$T M = \{(p, v) \mid p \in M, v \in T_p M\} = \coprod_{p \in M} T_p M.$$

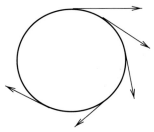

Figure 5.1. A particle moving on S^1: some of the velocity vectors are drawn. The collection of all possible combinations of position and velocity ought to assemble into a "tangent bundle". In this case we see that $S^1 \times \mathbf{R}^1$ would do, but in most instances it won't be as easy as this.

In the special case $M = \mathbf{R}^m$ we have a global chart (e.g., the identity chart), and so we have a global (not depending on the point p) identification of $T_p\mathbf{R}$ with the vector space $\mathbf{R}^m$ through the correspondence $[\gamma] \leftrightarrow \gamma'(0)$. Hence it is reasonable to say that $T\mathbf{R}^m$ can be identified with the product $\mathbf{R}^m \times \mathbf{R}^m$. Now, in general a manifold M is locally like $\mathbf{R}^m$, but the question is how this local information should be patched together to a global picture.

The tangent bundle is an example of an important class of objects called *vector bundles*. We start the discussion of vector bundles in general in this chapter, although our immediate applications will focus on the tangent bundle. We will pick up the glove in Chapter 6, where we discuss the algebraic properties of vector bundles, giving tools that eventually could have brought the reader to fascinating topics like the topological K-theory of Atiyah and Hirzebruch [1] which is an important tool in algebraic topology.

We first introduce topological vector bundles, and then see how transition functions, which are very similar to the chart transformations, allow us to coin what it means for a bundle to be smooth. An observation shows that the work of checking that something actually **is** a vector bundle can be significantly simplified, paving the way for the sleek definition of the tangent bundle in Definition 5.5.1. Using the same setup we easily get the cotangent bundle as well, see Section 5.6.

If you like electrons and want to start with an example instead of with the theory, you are invited to study Example 5.1.13 before reading on.

5.1 Topological Vector Bundles

Loosely speaking, a vector bundle is a collection of vector spaces parametrized in a locally controllable fashion by some space (Figure 5.2).

The easiest example is simply the product $X \times \mathbf{R}^n$, and we will have this as our local model (Figure 5.3).

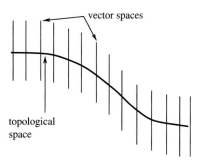

Figure 5.2. A vector bundle is a topological space to which a vector space is stuck at each point, and everything fitted continuously together.

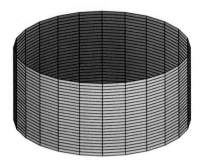

Figure 5.3. The product of a space and a Euclidean space is the local model for vector bundles. The cylinder $S^1 \times \mathbf{R}$ is an example.

Definition 5.1.1 An *n-dimensional (real topological) vector bundle* is a continuous map

$$\begin{array}{c} E \\ \pi \downarrow \\ X \end{array}$$

and a choice for each $p \in X$ of a real vector space structure of dimension n on the *fiber* $\pi^{-1}(p)$, such that for each $p \in X$

- there is an open set $U \subseteq X$ containing p and
- a homeomorphism $h \colon \pi^{-1}(U) \to U \times \mathbf{R}^n$ such that

$$\pi^{-1}(U) \xrightarrow{\quad h \quad} U \times \mathbf{R}^n$$
$$\pi|_{\pi^{-1}(U)} \searrow \qquad \swarrow \mathrm{pr}_U$$
$$U$$

commutes, and such that for every $q \in U$ the composite

$$h_q \colon \pi^{-1}(q) \xrightarrow{h|_{\pi^{-1}(q)}} \{q\} \times \mathbf{R}^n \xrightarrow{(q,t) \mapsto t} \mathbf{R}^n$$

is a vector space isomorphism.

Note 5.1.2 The map $\pi \colon E \to X$ is surjective since we have insisted that each fiber $\pi^{-1}(p)$ has a vector space structure and hence is not empty (it contains 0).

To distinguish the dimension of the fibers from the dimension of manifolds (X often is a manifold), we may also say that the vector bundle has *rank n*. A vector bundle of rank 1 is often referred to as a *line bundle*. It is not uncommon to mention just the total space E when referring to the vector bundle.

Example 5.1.3 The "unbounded Möbius band" (Figure 5.4) given by

$$\eta_1 = ([0, 1] \times \mathbf{R})/((0, p) \sim (1, -p))$$

defines a line bundle by projecting onto the first coordinate $\eta_1 \rightarrow [0, 1]/(0 \sim 1) \cong S^1$.

Restricting to an interval on the circle, we clearly see that it is homeomorphic to the product as shown in Figure 5.5.

This bundle is often referred to as the *tautological line bundle*. The reason for the name tautological line bundle is that, by Exercise 2.4.11, we know that $\mathbf{RP}^1$ and S^1 are diffeomorphic, and over $\mathbf{RP}^n$ we do have a "tautological line bundle" $\eta_n \rightarrow \mathbf{RP}^n$, where η_n is the space of pairs (L, v), where L is a line (one-dimensional linear subspace) in $\mathbf{R}^{n+1}$ and v a vector in L. The map is given by $(L, v) \mapsto L$. We will prove in Exercise 5.4.4 that this actually defines a vector bundle. The tautological line bundles are very important because they, in a precise sense, classify all line bundles. See Section 6.8 for some inadequate remarks.

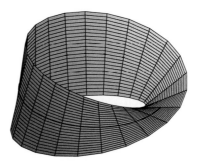

Figure 5.4.

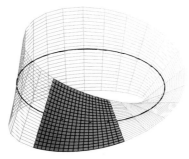

Figure 5.5.

Exercise 5.1.4   Consider the tautological line bundle (unbounded Möbius band)

$$\eta_1 \to S^1$$

from Example 5.1.3. Prove that there is no smooth map $f : \eta_1 \to \mathbf{R}$ such that the central circle $([0, 1] \times \{0\})/(0, 0) \sim (1, 0) \subseteq \eta_1$ is the inverse image of a regular value of f.

     More generally, show that there is no map $f : \eta_1 \to N$ for any manifold N such that the central circle is the inverse image of a regular value of f.

Definition 5.1.5   Given a rank-n topological vector bundle $\pi : E \to X$, we call

$E_q = \pi^{-1}(q)$ the *fiber* over $q \in X$,
E the *total space*, and
X the *base space* of the vector bundle.

     The existence of the (h, U)s is referred to as the *local trivialization* of the bundle ("the bundle is *locally trivial*"), and the (h, U)s are called *bundle charts*. A *bundle atlas* is a collection $\mathcal{B}$ of bundle charts that "covers" X, i.e., is such that

$$X = \bigcup_{(h,U) \in \mathcal{B}} U.$$

Note 5.1.6   Note the correspondence that the definition spells out between the homeomorphism h and the isomorphism h_q: for $r \in \pi^{-1}(U)$ we have

$$h(r) = (\pi(r), h_{\pi(r)}(r)).$$

Example 5.1.7   Given a topological space X, the projection onto the first factor

$$X \times \mathbf{R}^n$$

$$\mathrm{pr}_X \downarrow$$

$$X$$

is a rank-n topological vector bundle.

     This example is so totally uninteresting that we call it the *trivial bundle over X* (or, more descriptively, the *product bundle*). More generally, any vector bundle $\pi : E \to X$ with a bundle chart (h, X) is called trivial.

Definition 5.1.8   Let $\pi : E \to X$ be a vector bundle. A *section* to π is a continuous map $\sigma : X \to E$ such that $\pi\sigma(p) = p$ for all $p \in X$.

Example 5.1.9   Every vector bundle $\pi : E \to X$ has a section, namely the *zero section*, which is the map $\sigma_0 : X \to E$ that sends $p \in X$ to zero in the vector space $\pi^{-1}(p)$. As for any section, the map onto its image $X \to \sigma_0(X)$ is a

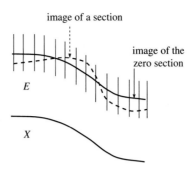

Figure 5.6.

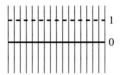

Figure 5.7. The trivial bundle has nonvanishing sections.

homeomorphism, and we will occasionally not distinguish between X and $\sigma_0(X)$ (we already did this when we talked informally about the unbounded Möbius band). See Figure 5.6.

Example 5.1.10   If $n > 0$, the trivial bundle $X \times \mathbf{R}^n \to X$ has *nonvanishing sections* as shown in Figure 5.7 (i.e., a section whose image does not intersect the zero section); for instance, if we choose any nonzero vector $v \in \mathbf{R}^n$, the section $p \mapsto (p, v)$ will do. The tautological line bundle $\eta_1 \to S^1$ (the unbounded Möbius band of Example 5.1.3), however, does not. This follows by the intermediate value theorem: a continuous function $f : [0, 1] \to \mathbf{R}$ with $f(0) = -f(1)$ must have a zero.

We have to specify the maps connecting the vector bundles. They come in two types, according to whether we allow the base space to change. The more general is specified in the following definition.

Definition 5.1.11   A *bundle morphism* from one bundle $\pi : E \to X$ to another $\pi' : E' \to X'$ is a pair of (continuous) maps

$$f : X \to X' \text{ and } \tilde{f} : E \to E'$$

such that

$$
\begin{array}{ccc}
E & \xrightarrow{\tilde{f}} & E' \\
\pi \downarrow & & \pi' \downarrow \\
X & \xrightarrow{f} & X'
\end{array}
$$

commutes, and such that each restriction to a fiber

$$\tilde{f}|_{\pi^{-1}(p)} \colon \pi^{-1}(p) \to (\pi')^{-1}(f(p))$$

is a linear map.

The composite of bundle morphisms is defined in the obvious way. For many purposes the most important bundle morphisms are those where the map on the base space is the identity given in the following definition.

Definition 5.1.12 Given a space X, a *bundle morphism over X* is a bundle morphism of the form

$$
\begin{array}{ccc}
E & \xrightarrow{\ \tilde{f}\ } & E' \\
\pi \downarrow & & \pi' \downarrow \\
X & =\!=\!= & X.
\end{array}
$$

An *isomorphism (over X)* of two vector bundles $\pi \colon E \to X$ and $\pi' \colon E' \to X$ over the same base space X is an invertible bundle morphism over X.

We will show in Lemma 5.3.12 that, essentially because a bijective linear map is an isomorphism, a bundle morphism over X is an isomorphism if and only if it is a bijection.

Note that a bundle is trivial if and only if it is isomorphic to a product bundle (see Exercise 5.1.7).

Bundles abound in applications. Here is a fairly simple one which demonstrates some important aspects.

Example 5.1.13 Consider two electrons of equal spin exposed to outside forces dictating that the electrons are equidistant from the origin in $\mathbf{R}^3$ (but not necessarily antipodal as in Section 1.2). How can we describe the space of possible configurations? Again, the two electrons are indistinguishable and cannot occupy the same point; the distance between the electrons does not contribute anything interesting and we'll ignore it. One way to record such a situation is to give the line in $\mathbf{R}^3$ which the electrons span and the center of mass. However, not every pair of a line and a vector gives a configuration: the center of mass is the point on the line closest to the origin (here I use the equidistance property).

A faithful description is thus to give a line through the origin $L \in \mathbf{RP}^2$ and a vector $v \in L^\perp$ (to the center of mass).

This is an example of a vector bundle: the base space is $\mathbf{RP}^2$ and each point in the total space is given by specifying a point $L \in \mathbf{RP}^2$ and a vector v in the fiber $L^\perp$.

Note that the vector space $L^\perp$ changes with L, so, although there are isomorphisms $L^\perp \cong \mathbf{R}^2$, these may obviously depend on L and we are not simply considering $\mathbf{RP}^2 \times \mathbf{R}^2$: there are twists.

Locally these twists can be undone. For instance, consider the open subset $U^0 \subseteq$ $\mathbf{RP}^2$. Saying that $L = [p] \in U^0$ (i.e., $p_0 \neq 0$) is the same as saying that the projection

$$h_L: L^\perp \to \mathbf{R}^2, \qquad h_L\left(\begin{bmatrix} v_0 \\ v_1 \\ v_2 \end{bmatrix}\right) = \begin{bmatrix} v_1 \\ v_2 \end{bmatrix}$$

is a linear isomorphism. The inverse of h_L is given by sending $[v_1 \, v_2]^T \in \mathbf{R}^2$ to $[v_0 \, v_1, v_2]^T \in L^\perp$, where $v_0 = -p_1 v_1/p_0 - p_2 v_2/p_0$. Consequently, the space of all configurations (L, v) with $L \in U^0$ may be identified with $U^0 \times \mathbf{R}^2$ via the local trivialization (aka "gauge") sending (L, v) to $h(L, v) = (L, h_L(v))$.

What happens when we look at two different local trivializations? For $L = [p] \in U^0 \cap U^1$ we have, in addition to h (at least) one more given by sending (L, v) to $g(L, v) = (L, g_L(v)) = (L, [v_0 \, v_2]^T)$. How do these compare? The obvious thing is to look at the composite (called the "bundle chart transformation")

$$(U^0 \cap U^1) \times \mathbf{R}^2 \xrightarrow{gh^{-1}} (U^0 \cap U^1) \times \mathbf{R}^2.$$

Tracing through the definitions we see that

$$gh^{-1}\left(L, \begin{bmatrix} v_1 \\ v_2 \end{bmatrix}\right) = \left(L, \begin{bmatrix} -p_1/p_0 & -p_2/p_0 \\ 0 & 1 \end{bmatrix} \begin{bmatrix} v_1 \\ v_2 \end{bmatrix}\right),$$

from which we learn two things. First, the L is not touched; and second, the fiber is moved by a linear isomorphism *depending smoothly* on $L = [p]$ – in this instance given by multiplying by the invertible (since neither p_0 nor p_1 is zero) matrix

$$A_{[p]} = \begin{bmatrix} -p_1/p_0 & -p_2/p_0 \\ 0 & 1 \end{bmatrix}.$$

If we insist that the line L is in the plane ("restrictions of bundles" will reappear in Section 6.1), i.e., $L \in \mathbf{RP}^1 \cong S^1$, things simplify slightly. The curve $\gamma: \mathbf{R} \to \mathbf{RP}^1, t \mapsto [\cos t \, \sin t]$ winds around $\mathbf{RP}^1$ once as t travels from 0 to π and intersects $U^0 \cap U^1$ in two disjoint opens: when $t \in (0, \pi/2)$ and when $t \in (\pi/2, \pi)$. We see that $\det A_{\gamma(t)} = -\cos t/\sin t$ is negative when $t \in (0, \pi/2)$ and positive when $t \in (\pi/2, \pi)$. This is a clear indication of an irreparable twist akin to the Möbius band, see Exercise 5.3.15. In fact, our vector bundle *is* fundamentally twisted and very different from $\mathbf{RP}^2 \times \mathbf{R}^2$.

5.2 Transition Functions

We will need to endow our bundles with smooth structures, and in order to do this we will use the same trick as we used to define manifolds: transport everything down to issues in Euclidean spaces. Given two overlapping bundle charts (h, U) and (g, V), on restricting to $\pi^{-1}(U \cap V)$ both define homeomorphisms

$$\pi^{-1}(U \cap V) \to (U \cap V) \times \mathbf{R}^n,$$

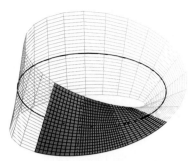

Figure 5.8. Two bundle charts. Restricting to their intersection, how do the two homeomorphisms to $(U \cap V) \times \mathbf{R}^n$ compare?

which we may compose to give homeomorphisms of $(U \cap V) \times \mathbf{R}^n$ with itself (Figure 5.8). If the base space is a smooth manifold, we may ask whether this map is smooth.

We need some names to talk about this construction.

Definition 5.2.1 Let $\pi \colon E \to X$ be a rank-n topological vector bundle, and let $\mathcal{B}$ be a bundle atlas. If $(h, U), (g, V) \in \mathcal{B}$ then

$$gh^{-1}|_{(U \cap V) \times \mathbf{R}^n} \colon (U \cap V) \times \mathbf{R}^n \to (U \cap V) \times \mathbf{R}^n$$

are called the *bundle chart transformations*. The restrictions to each fiber

$$g_q h_q^{-1} \colon \mathbf{R}^n \to \mathbf{R}^n$$

are linear isomorphisms (i.e., elements in $\mathrm{GL}_n(\mathbf{R})$) and the associated functions

$$U \cap V \to \mathrm{GL}_n(\mathbf{R}),$$
$$q \mapsto g_q h_q^{-1}$$

are called *transition functions*.

Again, visually bundle chart transformations are given by going up and down in

$$\pi^{-1}(U \cap V)$$

$h|_{\pi^{-1}(U \cap V)}$ $\qquad$ $g|_{\pi^{-1}(U \cap V)}$

$(U \cap V) \times \mathbf{R}^n$ $\qquad\qquad$ $(U \cap V) \times \mathbf{R}^n.$

The following lemma explains why giving the bundle chart transformations or the transition functions amounts to the same thing (and so it is excusable to confuse the two after a while; also, one should note that the terminology varies from author to author).

Lemma 5.2.2 *Let W be a topological space, and $f\colon W \to M_{m \times n}(\mathbf{R})$ a function. Then the associated function*

$$f_* \colon W \times \mathbf{R}^n \to \mathbf{R}^m,$$
$$(w, v) \mapsto f(w) \cdot v$$

is continuous if and only if f is. If W is a smooth manifold, then f_* is smooth if and only if f is.

Proof. Note that f_* is the composite

$$W \times \mathbf{R}^n \xrightarrow{\ f \times \mathrm{id}\ } M_{m \times n}(\mathbf{R}) \times \mathbf{R}^n \xrightarrow{\ e\ } \mathbf{R}^m,$$

where $e(A, v) = A \cdot v$. Since e is smooth, it follows that, if f is continuous or smooth, then so is f_*.

Conversely, considered as a matrix, we have that

$$[f(w)] = \left[f_*(w, e_1), \dots, f_*(w, e_n) \right].$$

If f_* is continuous (or smooth), then we see that each column of $[f(w)]$ depends continuously (or smoothly) on w, and so f is continuous (or smooth). ∎

This lemma will allow us to rephrase the definition of vector bundles in terms of so-called pre-vector bundles (see Definition 5.4.1), which reduces the amount of checking needed to see that some example actually is a vector bundle. Also, the lemma has the following corollary which will allow us to alternate whether we want to check smoothness assumptions on bundle chart transformations or on transition functions.

Corollary 5.2.3 *Let $E \to M$ be a vector bundle over a smooth manifold and let $\mathcal{B}$ be a bundle atlas. A bundle chart transformation in $\mathcal{B}$ is smooth if and only if the associated transition function is smooth.*

A nice formulation of the contents of Lemma 5.2.2 is that we have a bijection from the set of continuous functions $W \to M_{m \times n}\mathbf{R}$ to the set of bundle morphisms

$$W \times \mathbf{R}^n \xrightarrow{\hspace{3cm}} W \times \mathbf{R}^m$$

with pr_W and pr_W mapping down to W.

by sending the function $g \colon W \to M_{m \times n}\mathbf{R}$ to the function

$$G(g) \colon W \times \mathbf{R}^n \xrightarrow{\ (p,v) \mapsto (p, g(p) \cdot v)\ } W \times \mathbf{R}^m.$$

Furthermore, if W is a smooth manifold, then g is smooth if and only if $G(g)$ is smooth.

Exercise 5.2.4 Show that any vector bundle $E \to [0, 1]$ is trivial.

Exercise 5.2.5 Show that any line bundle (rank-1 vector bundle) $E \to S^1$ is either trivial or isomorphic to the tautological line bundle η_1. Show the analogous statement for rank n vector bundles over S^1.

5.3 Smooth Vector Bundles

Definition 5.3.1 Let M be a smooth manifold, and let $\pi: E \to M$ be a vector bundle. A bundle atlas is said to be *smooth* if all the transition functions are smooth.

Note 5.3.2 On spelling the differentiability out in full detail we get the following. Let $(M, \mathcal{A})$ be a smooth n-dimensional manifold, $\pi: E \to M$ a rank-k vector bundle and $\mathcal{B}$ a bundle atlas. Then $\mathcal{B}$ is smooth if, for every pair of bundle charts $(h_1, U_1), (h_2, U_2) \in \mathcal{B}$ and every pair of charts $(x_1, V_1), (x_2, V_2) \in \mathcal{A}$, the composite going up, over and across

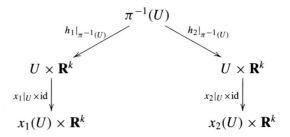

is a smooth function in $\mathbf{R}^{n+k}$, where $U = U_1 \cap U_2 \cap V_1 \cap V_2$.

Example 5.3.3 If M is a smooth manifold, then the trivial bundle is a smooth vector bundle in an obvious manner.

Example 5.3.4 The tautological line bundle (unbounded Möbius strip of Example 5.1.3) $\eta_1 \to S^1$ is a smooth vector bundle. As a matter of fact, the trivial bundle and the tautological line bundle are, up to isomorphism, the only line bundles over the circle (see Exercise 5.3.15 for the smooth case or Exercise 5.2.5 for the topological case).

Note 5.3.5 Just as for atlases of manifolds, we have a notion of a *maximal (smooth) bundle atlas*, and with each smooth atlas we may associate a unique maximal one in exactly the same way as before.

Definition 5.3.6 A *smooth vector bundle* is a vector bundle equipped with a maximal smooth bundle atlas.

We will often suppress the bundle atlas from the notation, so, if the maximal atlas is clear from the context, a smooth vector bundle $(\pi: E \to M, \mathcal{B})$ will be written simply $\pi: E \to M$ (or, even worse, E).

Definition 5.3.7 A smooth vector bundle $(\pi: E \to M, \mathcal{B})$ is *trivial* if its (maximal smooth) atlas $\mathcal{B}$ contains a chart (h, M) with domain all of M.

Lemma 5.3.8 *The total space E of a smooth vector bundle $(\pi: E \to M, \mathcal{B})$ has a natural smooth structure, and π is a smooth map.*

Proof. Let M be n-dimensional with atlas $\mathcal{A}$, and let the bundle be of rank (dimension) k. Then the diagram in Note 5.3.2 shows that E is a smooth $(n + k)$-dimensional manifold. That π is smooth is the same as claiming that all the up, over and across composites

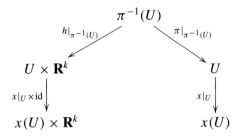

are smooth, where $(x, V) \in \mathcal{A}$, $(h, W) \in \mathcal{B}$ and $U = V \cap W$. But

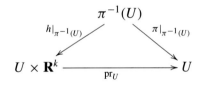

commutes, so the composite is simply the projection $\mathrm{pr}_{x(U)}: x(U) \times \mathbf{R}^k \to x(U)$, which is smooth. ∎

Note 5.3.9 As expected, the proof shows that $\pi: E \to M$ locally looks like the projection

$$\mathbf{R}^n \times \mathbf{R}^k \to \mathbf{R}^n.$$

Exercise 5.3.10 Show that the following is an equivalent definition to the one we have given. A smooth rank-k vector bundle is a smooth map $\pi: E \to M$ together with a vector space structure on each fiber such that for each $p \in M$ there is an open $U \subseteq M$ containing p and a diffeomorphism $h: \pi^{-1}(U) \to U \times \mathbf{R}^k$ which is linear on each fiber and with $\mathrm{pr}_U h(e) = \pi(e)$.

Definition 5.3.11   A *smooth bundle morphism* is a bundle morphism in the sense of Definition 5.1.11

$$
\begin{array}{ccc}
E & \xrightarrow{\tilde{f}} & E' \\
\pi \downarrow & & \pi' \downarrow \\
M & \xrightarrow{f} & M'
\end{array}
$$

from a smooth vector bundle to another such that $\tilde{f}$ and f are smooth.

An *isomorphism* of two smooth vector bundles

$$
\pi : E \to M \text{ and } \pi' : E' \to M
$$

over the same base space M is an invertible smooth bundle morphism over the identity on M:

$$
\begin{array}{ccc}
E & \xrightarrow{\tilde{f}} & E' \\
\pi \downarrow & & \pi' \downarrow \\
M & = & M .
\end{array}
$$

The term "invertible smooth bundle morphism" signifies that the inverse is a smooth bundle morphism. However, checking whether a bundle morphism is an isomorphism reduces to checking that it is a bijection.

Lemma 5.3.12   *Let*

$$
\begin{array}{ccc}
E & \xrightarrow{\tilde{f}} & E' \\
\pi \downarrow & & \pi' \downarrow \\
M & = & M
\end{array}
$$

be a smooth (or continuous) bundle morphism. If $\tilde{f}$ is bijective, then it is a smooth (or continuous) isomorphism.

Proof.   That $\tilde{f}$ is bijective means that it is a bijective linear map on every fiber, or, in other words, a vector space isomorphism on every fiber. Choose charts (h, U) in E and (h', U) in E' around $p \in U \subseteq M$ (one may choose the Us to be the same). Then

$$
h' \tilde{f} h^{-1} : U \times \mathbf{R}^n \to U \times \mathbf{R}^n
$$

is of the form $(u, v) \mapsto (u, \alpha_u v)$, where $\alpha_u \in \mathrm{GL}_n(\mathbf{R})$ depends smoothly (or continuously) on $u \in U$. But by Cramer's rule $(\alpha_u)^{-1}$ depends smoothly on α_u, and so the inverse

$$
\left(h' \tilde{f} h^{-1} \right)^{-1} : U \times \mathbf{R}^n \to U \times \mathbf{R}^n, \qquad (u, v) \mapsto (u, (\alpha_u)^{-1} v)
$$

is smooth (or continuous), proving that the inverse of $\tilde{f}$ is smooth (or continuous). ∎

Exercise 5.3.13

Let a be a real number and $E \to X$ a bundle. Show that multiplication by a in each fiber gives a bundle morphism

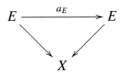

which is an isomorphism if and only if $a \neq 0$. If $E \to X$ is a smooth vector bundle, then a_E is smooth too.

Exercise 5.3.14

Show that any smooth vector bundle $E \to [0, 1]$ is trivial (smooth on the boundary means what you think it does: either don't worry or check up on Note 2.3.17 on smooth manifolds with boundary).

Exercise 5.3.15

Show that any smooth line bundle over S^1 is either trivial or isomorphic to the tautological line bundle of Example 5.1.3. Show the analogous statement for rank-n smooth vector bundles over S^1.

5.4 Pre-vector Bundles

A smooth or topological vector bundle is a very structured object, and much of its structure is intertwined very closely. There is a sneaky way out of having to check topological properties all the time. As a matter of fact, the topology is determined by some of the other structure as soon as the claim that it is a vector bundle is made: specifying the topology on the total space is redundant!

Definition 5.4.1   A *pre-vector bundle* of dimension (or rank) n is

a set E (*total space*)
a topological space X (*base space*)
a function $\pi : E \to X$
a vector space structure on the fiber $\pi^{-1}(q)$ for each $q \in X$
a *pre-bundle atlas* $\mathcal{B}$, i.e., a set $\mathcal{B}$ of pairs (h, U) with

$\qquad$ U an open subset of X and
$\qquad$ h a bijective function

$$\pi^{-1}(U) \xrightarrow{\;e \mapsto h(e) = (\pi(e), h_{\pi(e)}(e))\;} U \times \mathbf{R}^n$$

$\qquad$ which is linear on each fiber,

such that

$\qquad$ $\mathcal{B}$ covers X and
$\qquad$ the transition functions are continuous.

That $\mathcal{B}$ covers X means that $X = \bigcup_{(h,U)\in\mathcal{B}} U$; that h is linear on each fiber means that $h_q \colon \pi^{-1}(q) \to \mathbf{R}^n$ is linear for each $q \in X$; and that the transition functions of $\mathcal{B}$ are continuous means that, if $(h, U), (h', U') \in \mathcal{B}$, then

$$U \cap U' \to \mathrm{GL}_n(\mathbf{R}), \quad q \mapsto h'_q h_q^{-1}$$

is continuous.

Definition 5.4.2 A *smooth pre-vector bundle* is a pre-vector bundle where the base space is a smooth manifold and the transition functions are smooth.

Lemma 5.4.3 *Given a pre-vector bundle, there is a unique vector bundle with underlying pre-vector bundle the given one. The same statement holds for the smooth case.*

Proof. Let $(\pi \colon E \to X, \mathcal{B})$ be a pre-vector bundle. We must equip E with a topology such that π is continuous and the bijections in the bundle atlas are homeomorphisms. The smooth case follows then immediately from the continuous case.

We must have that, if $(h, U) \in \mathcal{B}$, then $\pi^{-1}(U)$ is an open set in E (for π to be continuous). The family of open sets $\{\pi^{-1}(U)\}_{U \subseteq X \text{ open}}$ covers E, so we need to know only what the open subsets of $\pi^{-1}(U)$ are, but this follows by the requirement that the bijection h should be a homeomorphism. That is, $V \subseteq \pi^{-1}(U)$ is open if $V = h^{-1}(V')$ for some open $V' \subseteq U \times \mathbf{R}^k$ (Figure 5.9). Ultimately, we get that

$$\left\{ h^{-1}(V_1 \times V_2) \,\middle|\, (h, U) \in \mathcal{B}, \begin{array}{l} V_1 \quad \text{open in } U, \\ V_2 \quad \text{open in } \mathbf{R}^k \end{array} \right\}$$

is a basis for the topology on E. ∎

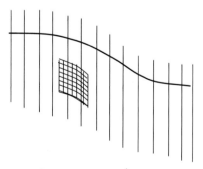

Figure 5.9. A typical open set in $\pi^{-1}(U)$ gotten as h^{-1} of the product of an open set in U and an open set in $\mathbf{R}^k$

Exercise 5.4.4

Let
$$\eta_n = \left\{ ([p], \lambda p) \in \mathbf{RP}^n \times \mathbf{R}^{n+1} \mid p \in S^n, \lambda \in \mathbf{R} \right\}.$$

Show that the projection

$$\eta_n \to \mathbf{RP}^n,$$
$$([p], v) \mapsto [p]$$

defines a non-trivial smooth vector bundle, called the *tautological line bundle*. Be careful that your proposed bundle charts are well defined.

Exercise 5.4.5

Let $p \in \mathbf{RP}^n$ and $X = \mathbf{RP}^n \setminus \{p\}$. Show that X is diffeomorphic to the total space η_{n-1} of the tautological line bundle in Exercise 5.4.4.

Exercise 5.4.6

You are given two pre-vector bundles over a common base space and a function $\tilde{f} \colon E \to E'$ between their total spaces. Spell out the condition for this to define a bundle morphism.

Note 5.4.7 Don't look at this until you've done Exercise 5.4.4: as always, you should do it without looking at the hint. However, it is handy to have a reference for a bundle atlas for the tautological line bundle: letting $\{(x^k, U^k)\}$ be the standard atlas for $\mathbf{RP}^n$, you get a bundle chart $h_k \colon \pi^{-1} U^k \to U^k \times \mathbf{R}$ (where $\pi \colon \eta_n \to \mathbf{RP}^n$ is the projection) by sending $([p], (v_0, \dots, v_n))$ to $([p], v_k)$ (the inverse is given by $h_k^{-1}([p], t) = ([p], (t/p_k)p))$.

5.5 The Tangent Bundle

We define the tangent bundle as follows.

Definition 5.5.1 Let $(M, \mathcal{A})$ be a smooth m-dimensional manifold. The *tangent bundle* of M is defined by the following smooth pre-vector bundle:

$TM = \coprod_{p \in M} T_p M$ (total space)
M (base space)
$\pi \colon TM \to M$ sends $T_p M$ to p
the pre-vector bundle atlas

$$\mathcal{B}_\mathcal{A} = \{(h_x, U) \mid (x, U) \in \mathcal{A}\}$$

where h_x is given by

$$h_x \colon \pi^{-1}(U) \to U \times \mathbf{R}^m,$$
$$[\gamma] \mapsto (\gamma(0), (x\gamma)'(0)).$$

Note 5.5.2 We recognize the local trivialization $h_x \colon \pi^{-1}(U) \to U \times \mathbf{R}^m$ given in the definition of the tangent bundle: on the fiber above a point $p \in U$ it is nothing but the isomorphism $A_x \colon T_p M \cong \mathbf{R}^m$ of Lemma 3.3.10. We see that the tangent bundle is smooth (as claimed), since the transition functions are given by multiplication by the Jacobi matrix: $h_y h_x^{-1}(p, v) = (p, D(yx^{-1})(x(p)) \cdot v)$.

Strictly speaking, an element in TM is a pair $(p, [\gamma])$ with $p \in M$ and $[\gamma] \in T_p M$, but we'll often abbreviate this to $[\gamma]$, letting γ bear the burden of remembering $p = \gamma(0)$.

Note 5.5.3   Since the tangent bundle is a smooth vector bundle, the total space TM is a smooth $2m$-dimensional manifold. To be explicit, its atlas is gotten from the smooth atlas on M as follows.

If (x, U) is a chart on M,

$$\pi^{-1}(U) \xrightarrow{\ h_x\ } U \times \mathbf{R}^m \xrightarrow{\ x \times \mathrm{id}\ } x(U) \times \mathbf{R}^m$$

$$[\gamma] \longmapsto (x\gamma(0), (x\gamma)'(0))$$

is a homeomorphism to an open subset of $\mathbf{R}^m \times \mathbf{R}^m$. It is convenient to have an explicit formula for the inverse (c.f. Lemma 3.3.10): it sends $(q, v) \in x(U) \times \mathbf{R}^m$ to the tangent vector $[x^{-1}\gamma_{(q,v)}]$ in the fiber $T_{x^{-1}(q)}M$, where $\gamma_{(q,v)} \colon (\mathbf{R}, 0) \to (\mathbf{R}^m, q)$ is the germ defined by the straight line sending t to $q + tv$.

Lemma 5.5.4   *Let* $f \colon (M, \mathcal{A}_M) \to (N, \mathcal{A}_N)$ *be a smooth map. Then*

$$[\gamma] \mapsto Tf[\gamma] = [f\gamma]$$

defines a smooth bundle morphism

$$
\begin{array}{ccc}
TM & \xrightarrow{\ Tf\ } & TN \\
{\scriptstyle \pi_M}\downarrow & & {\scriptstyle \pi_N}\downarrow \\
M & \xrightarrow{\ f\ } & N.
\end{array}
$$

Proof.   Since $Tf|_{\pi^{-1}(p)} = T_p f$ we have linearity on the fibers, and we are left with showing that Tf is a smooth map. Let $(x, U) \in \mathcal{A}_M$ and $(y, V) \in \mathcal{A}_N$. We have to show that up, across and down in

$$
\begin{array}{ccc}
\pi_M^{-1}(W) & \xrightarrow{\ Tf|\ } & \pi_N^{-1}(V) \\
{\scriptstyle h_x|_W}\downarrow & & {\scriptstyle h_y}\downarrow \\
W \times \mathbf{R}^m & & V \times \mathbf{R}^n \\
{\scriptstyle x|_W \times \mathrm{id}}\downarrow & & {\scriptstyle y \times \mathrm{id}}\downarrow \\
x(W) \times \mathbf{R}^m & & y(V) \times \mathbf{R}^n
\end{array}
$$

is smooth, where $W = U \cap f^{-1}(V)$ and $Tf|$ is Tf restricted to $\pi_M^{-1}(W)$. This composite sends $(q, v) \in x(W) \times \mathbf{R}^m$ to $[x^{-1}\gamma_{(q,v)}] \in \pi_M^{-1}(W)$ to $[fx^{-1}\gamma_{(q,v)}] \in$

$\pi_N^{-1}(V)$ and finally to $(yfx^{-1}\gamma_{(q,v)}(0), (yfx^{-1}\gamma_{(q,v)})'(0)) \in y(V) \times \mathbf{R}^n$, which is equal to

$$(yfx^{-1}(q), D(yfx^{-1})(q) \cdot v)$$

by the chain rule. Since yfx^{-1} is a smooth function, this is a smooth function too. ∎

Lemma 5.5.5 *If $f: M \to N$ and $g: N \to L$ are smooth, then*

$$TgTf = T(gf).$$

Proof. It is the chain rule in Lemma 3.3.6 (made pleasant since the notation no longer has to specify over which point in your manifold you are). ∎

Note 5.5.6 The tangent space of $\mathbf{R}^n$ is trivial, since the identity chart induces a bundle chart

$$h_{\mathrm{id}}: T\mathbf{R}^n \to \mathbf{R}^n \times \mathbf{R}^n,$$
$$[\gamma] \mapsto (\gamma(0), \gamma'(0)).$$

Definition 5.5.7 A manifold is often said to be *parallelizable* if its tangent bundle is trivial.

Example 5.5.8 The circle is parallelizable. This is so since the map

$$S^1 \times T_1 S^1 \to TS^1,$$
$$(e^{i\theta}, [\gamma]) \mapsto [e^{i\theta} \cdot \gamma]$$

is both a diffeomorphism and linear on each fiber (here $(e^{i\theta} \cdot \gamma)(t) = e^{i\theta} \cdot \gamma(t)$) with inverse $[\gamma] \mapsto (\gamma(0), [\gamma(0)^{-1}\gamma])$.

Exercise 5.5.9 Show that the 3-sphere S^3 is parallelizable.

Exercise 5.5.10 Show that all Lie groups are parallelizable. (A Lie group is a manifold with a smooth associative multiplication, with a unit and all inverses: skip this exercise if this sounds too alien to you.)

Example 5.5.11 Let

$$E = \{(p, v) \in \mathbf{R}^{n+1} \times \mathbf{R}^{n+1} \mid |p| = 1, p \cdot v = 0\}.$$

Then

$$TS^n \to E, \qquad [\gamma] \mapsto (\gamma(0), \gamma'(0))$$

is a homeomorphism. The inverse sends $(p, v) \in E$ to the tangent vector $[t \mapsto (p + tv)/|p + tv|] \in T_p S^n$. See Figures 5.10 and 5.11.

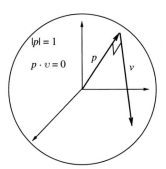

Figure 5.10. A point in the tangent space of S^2 may be represented by a unit vector p together with an arbitrary vector v perpendicular to p.

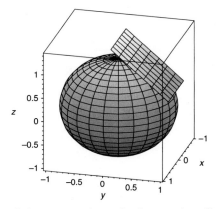

Figure 5.11. We can't draw all the tangent planes simultaneously to illustrate the tangent bundle of S^2. The description we give is in $\mathbf{R}^6$.

More generally we have the following fact.

Lemma 5.5.12 *Let $f\colon M \to N$ be an imbedding. Then $Tf\colon TM \to TN$ is an imbedding.*

Proof. We may assume that f is the inclusion of an m-dimensional submanifold in an $(m+k)$-dimensional manifold (the diffeomorphism part is taken care of by the chain rule which implies that if f is a diffeomorphism then Tf is a diffeomorphism with inverse $T(f^{-1})$).

In this case, TM is a subset of TN with inclusion Tf, and we must display charts for TN displaying TM as a $2m$-dimensional submanifold. Let $y\colon V \to V'$ be a chart on N such that $y(V \cap M) = V' \cap (\mathbf{R}^m \times \{0\})$. For notational simplicity, we identify the subspace $\mathbf{R}^m \times \{0\}$ of $\mathbf{R}^{m+k}$ with $\mathbf{R}^m$. Since curves in $\mathbf{R}^m$ have derivatives in $\mathbf{R}^m$ we get a commuting diagram

$$T(V \cap M) \xrightarrow[\ T_{y_{V \cap M}}\ ]{\cong} T(V' \cap \mathbf{R}^m) \xrightarrow{\cong} (V' \cap \mathbf{R}^m) \times \mathbf{R}^m$$

$$
\begin{array}{ccc}
\Big\downarrow {\scriptstyle T f_{V \cap M}} & & \Big\downarrow {\scriptstyle \text{inclusion}}
\end{array}
$$

$$TV \xrightarrow[\ T_y\ ]{\cong} TV' \xrightarrow{\cong} V' \times \mathbf{R}^{m+k},$$

where the rightmost diffeomorphisms are the standard identifications sending a tangent vector $[\gamma]$ to $(\gamma(0), \gamma'(0))$. The horizontal composites are just the charts for TM and TN associated with the chart (y, V), and since $TV \cap TM = T(V \cap M)$ the diagram shows that the charts are of the required form. ∎

Corollary 5.5.13 *If $M \subseteq \mathbf{R}^N$ is the inclusion of a smooth submanifold of a Euclidean space, then*

$$TM \cong \left\{ (p, v) \in \mathbf{M} \times \mathbf{R}^N \ \middle|\ \begin{array}{l} v = \gamma'(0) \text{ for some germ} \\ \bar{\gamma} \colon (\mathbf{R}, 0) \to (M, p) \end{array} \right\} \subseteq \mathbf{R}^N \times \mathbf{R}^N \cong T\mathbf{R}^N$$

(the derivation of γ happens in $\mathbf{R}^N$).

Exercise 5.5.14 There is an even groovier description of TS^n: prove that

$$E = \left\{ (z_0, \ldots, z_n) \in \mathbf{C}^{n+1} \ \middle|\ \sum_{i=0}^{n} z^2 = 1 \right\}$$

is the total space in a bundle isomorphic to TS^n.

Definition 5.5.15 Let M be a smooth manifold. A *vector field* on M is a smooth section in the tangent bundle, i.e., a smooth map $\sigma \colon M \to TM$ such that the composite $\pi_M \sigma \colon M \to TM \to M$ is the identity.

Exercise 5.5.16 Let M be a smooth manifold and $\mathcal{X}(M)$ the set of vector fields on M. Using the vector space structure on the tangent spaces, give $\mathcal{X}(M)$ the structure of a vector space. If you know the language, extend this to a $\mathcal{C}^\infty(M)$-module structure, where $\mathcal{C}^\infty(M)$ is the ring of smooth functions $M \to \mathbf{R}$.

Exercise 5.5.17 Let M be an n-dimensional parallelizable manifold (i.e., the tangent bundle is trivial). Give an isomorphism $\mathcal{X}(M) \cong \mathcal{C}^\infty(M, \mathbf{R}^n) \cong \mathcal{C}^\infty(M)^{\times n}$ from the space of vector fields to n times the space of smooth real functions.

Exercise 5.5.18 Give an isomorphism $\mathcal{X}(S^n) \cong \{\sigma \in \mathcal{C}^\infty(S^n, \mathbf{R}^{n+1}) \mid p \cdot \sigma(p) = 0\}$ and an isomorphism

$$\mathcal{X}(S^n) \times \mathcal{C}^\infty(S^n) \cong \mathcal{C}^\infty(S^n)^{\times(n+1)}.$$

Exercise 5.5.19 Prove that the projection $S^n \to \mathbf{R}P^n$ gives an isomorphism

$$T\mathbf{R}P^n \cong \{(p, v) \in S^n \times \mathbf{R}^{n+1} \mid p \cdot v = 0\}/(p, v) \sim (-p, -v).$$

Exercise 5.5.20 Consider the inclusion $O(n) \subseteq M_n(\mathbf{R})$, giving a description of the tangent bundle of $O(n)$ along the lines of Corollary 5.5.13. Show that under the isomorphism

$$TM_n(\mathbf{R}) \cong M_n(\mathbf{R}) \times M_n(\mathbf{R}), \qquad [\gamma] \leftrightarrows (\gamma(0), \gamma'(0))$$

the tangent bundle of $O(n)$ corresponds to the projection on the first factor

$$E = \{(g, A) \in O(n) \times M_n(\mathbf{R}) \mid A^{\mathrm{T}} = -g^{\mathrm{T}} A g^{\mathrm{T}}\} \to O(n).$$

This also shows that $O(n)$ is parallelizable (which we knew already by Exercise 5.5.10, since $O(n)$ is a Lie group), since we get a bundle isomorphism induced by

$$E \to O(n) \times \mathrm{Skew}(n), \qquad g, A) \mapsto (g, g^{-1}A),$$

where $\mathrm{Skew}(n) = \{B \in M_n(\mathbf{R}) \mid B^{\mathrm{T}} = -B\}$ (a matrix B satisfying $B^{\mathrm{T}} = -B$ is called a *skew matrix*).

As we will see in Chapter 7, vector fields are closely related to differential equations. It is often of essence to know whether a manifold M supports *non-vanishing* vector fields, i.e., a vector field $s \colon M \to TM$ such that $s(p) \neq 0$ for all $p \in M$.

Example 5.5.21 The circle has nonvanishing vector fields (Figure 5.12). Let $[\gamma] \neq 0 \in T_1 S^1$, then

$$S^1 \to TS^1, \qquad e^{i\theta} \mapsto [e^{i\theta} \cdot \gamma]$$

is a vector field (since $e^{i\theta} \cdot \gamma(0) = e^{i\theta} \cdot 1$) and does not intersect the zero section since (viewed as a vector in $\mathbf{C}$)

$$|(e^{i\theta} \cdot \gamma)'(0)| = |e^{i\theta} \cdot \gamma'(0)| = |\gamma'(0)| \neq 0.$$

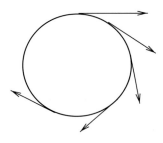

Figure 5.12. The vector field spins around the circle with constant speed.

This is the same construction as the one we used to show that S^1 was parallelizable. This is a general argument: an n-dimensional manifold with n linearly independent vector fields has a trivial tangent bundle, and conversely.

Exercise 5.5.22 Construct three vector fields on S^3 that are linearly independent in all tangent spaces.

Exercise 5.5.23 Prove that $T(M \times N) \cong TM \times TN$.

Example 5.5.24 We have just seen that S^1 and S^3 (if you did the exercise) both have nonvanishing vector fields. It is a hard fact that S^2 does not: "you can't comb the hair on a sphere".

This has the practical consequence that, when you want to confine the plasma in a fusion reactor by means of magnetic fields, you can't choose to let the plasma be in the interior of a sphere (or anything homeomorphic to it). At each point on the surface bounding the region occupied by the plasma, the component of the magnetic field parallel to the surface must be nonzero, or the plasma will leak out (if you remember your physics, there once was a formula saying something like $F = qv \times B$, where q is the charge of the particle, v its velocity and B the magnetic field: hence any particle moving nonparallel to the magnetic field will be deflected).

This problem is solved by letting the plasma stay inside a torus $S^1 \times S^1$ which does have nonvanishing vector fields (since S^1 has by virtue of Example 5.5.21, and since $T(S^1 \times S^1) \cong TS^1 \times TS^1$ by Exercise 5.5.23).

Although there are no nonvanishing vector fields on S^2, there are certainly interesting ones that have only a few zeros. For instance "rotation around an axis" will give you a vector field with only two zeros. The "magnetic dipole" defines a vector field on S^2 with just one zero (Figure 5.13).

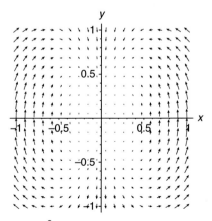

Figure 5.13. A magnetic dipole on S^2, seen by stereographic projection in a neighborhood of the only zero.

It turns out that, in the classification of orientable surfaces mentioned in Note 2.4.22, the torus is the only one to support a nonvanishing vector field (this has strangely to do with the Euler characteristic "$V - E + F$" of Exercise 1.5.4). As a spin-off, since we know that Lie groups are parallelizable and hence support nonvanishing vector fields, this gives that the torus is the only orientable surface supporting a Lie group structure.

Exercise 5.5.25   Let M be an n-dimensional smooth manifold. For $p \in M$, let E_p be the set of germs $\bar{\sigma} : (\mathbf{R}^2, 0) \to (M, p)$ modulo the equivalence relation that $\bar{\sigma}_1 \simeq \bar{\sigma}_2$ if for any chart (x, U) with $p \in U$ we have that

$$D_1(x\sigma_1)(0) = D_1(x\sigma_2)(0), \quad D_2(x\sigma_1)(0) = D_2(x\sigma_2)(0) \text{ and}$$
$$D_1 D_2(x\sigma_1)(0) = D_1 D_2(x\sigma_2)(0).$$

Let $E = \coprod_{p \in M} E_p$ and consider the projection $E \to TM$ sending $[\sigma]$ to $[t \mapsto \sigma(0, t)]$. Show that the assignment $E \to T(TM)$ sending $[\sigma]$ to $[s \mapsto [t \mapsto \sigma(s, t)]]$ is a well-defined bijection, making $E \to TM$ a smooth vector bundle isomorphic to the tangent bundle of TM.

5.6 The Cotangent Bundle[1]

Let M be a smooth m-dimensional manifold. Recall the definition of the cotangent spaces from Section 3.4, more precisely Definition 3.4.1. We will show that the cotangent spaces join to form a bundle, the *cotangent bundle* T^*M, by showing that they define a pre-vector bundle.

Let the total space T^*M be the set

$$T^*M = \coprod_{p \in M} T_p^*M = \{(p, d\phi) \mid p \in M, d\phi \in T_p^*M\}$$

and $\pi : T^*M \to M$ be the projection sending $(p, d\phi)$ to p. For a smooth chart (x, U) we have a bundle chart

$$h_x : \pi^{-1}(U) = T^*U \to U \times (\mathbf{R}^n)^*$$

gotten by sending $(p, d\phi)$ to $(p, D(\phi x^{-1})(x(p))\cdot)$. To get it in exactly the form of Definition 5.1.1 we should choose an isomorphism $\mathrm{Hom}_\mathbf{R}(\mathbf{R}^m, \mathbf{R}) = (\mathbf{R}^m)^* \cong \mathbf{R}^m$ once and for all (e.g., transposing vectors), but it is convenient to postpone this translation as long as possible.

By the discussion in Section 3.4, h_x induces a linear isomorphism $\pi^{-1}(p) = T_p^*M \cong \{p\} \times (\mathbf{R}^m)^*$ in each fiber. If (y, V) is another chart, the transition function is given by sending $p \in U \cap V$ to the linear isomorphism $(\mathbf{R}^m)^* \to (\mathbf{R}^m)^*$ induced by the linear isomorphism $\mathbf{R}^m \to \mathbf{R}^m$ given by multiplication by the Jacobi matrix

---

[1] If you did not read about the cotangent space in Section 3.4, you should skip this section.

$D(yx^{-1})(x(p))$. Since the Jacobi matrix $D(yx^{-1})(x(p))$ varies smoothly with p, we have shown that

$$T^*M \to M$$

is a smooth (pre-)vector bundle, the *cotangent bundle*.

Exercise 5.6.1 Go through the details in the above discussion.

Definition 5.6.2 If M is a smooth manifold, a *one-form* is a smooth section of the cotangent bundle $T^*M \to M$.

Example 5.6.3 Let $f \colon M \to \mathbf{R}$ be a smooth function. Recall the differential map $d \colon \mathcal{O}_{M,p} \to T_p^*M$ given by sending a function germ $\bar{\phi}$ to the cotangent vector represented by the germ of $q \mapsto \phi(q) - \phi(p)$. Correspondingly, we write $d_p f \in T_p^*M$ for the cotangent vector represented by $q \mapsto f(q) - f(p)$. Then the assignment $p \mapsto (p, d_p f) \in T^*M$ is a one-form, and we simply write

$$df \colon M \to T^*M.$$

We call this one-form df the *differential* of f.

Here's a concrete example. If $f \colon \mathbf{R} \to \mathbf{R}$ is given by $f(s) = s^2$, then under the trivialization $T^*\mathbf{R} \cong \mathbf{R} \times \mathbf{R}$ induced by the identity chart (i.e., given by $(s, d\phi) \mapsto (s, \phi'(s))$), df corresponds to $s \mapsto (s, 2s)$. Classically the identity chart could also be called "s", making the expression $d_s f = 2s\, ds$ meaningful (and admittedly more palatable than "$2s\, d(\mathrm{id}_\mathbf{R})$").

To signify that the differential df of Example 5.6.3 is just the beginning in a series of important vector spaces, let $\Omega^0(M) = \mathcal{C}^\infty(M, \mathbf{R})$ and let $\Omega^1(M)$ be the vector space of all one-forms on M. The differential is then a map

$$d \colon \Omega^0(M) \to \Omega^1(M).$$

Even though the differential as a map to each individual cotangent space $d \colon \mathcal{O}_{M,p} \to T_p^*M$ was surjective, this is not the case for $d \colon \Omega^0(M) \to \Omega^1(M)$. In fact, the one-forms in the image of d are the ones that are referred to as "exact". (This is classical notation coming from differential equations, the other relevant notion being "closed". It is the wee beginning of the study of the shapes of spaces through cohomological methods.)

Example 5.6.4 If $x_1, x_2 \colon S^1 \subseteq \mathbf{R}^2 \to \mathbf{R}$ are the projections to the first and second coordinate, respectively, then one can show that

$$x_1\, dx_2 - x_2\, dx_1$$

is a one-form that is not exact, a phenomenon related to the fact that the circle is not simply connected. As a matter of fact, the quotient $H^1(S^1) = \Omega^1(S^1)/d(\Omega^0(S^1))$ (which is known as the *first de Rham cohomology group* of the circle) is a one-dimensional real vector space, and so the image of the non-exact one-form displayed above generates $H^1(S^1)$.

Example 5.6.5 In physics the total space of the cotangent bundle is referred to as the *phase space*. If the manifold M is the collection of all possible positions of the physical system, the phase space T^*M is the collection of all positions and momenta. For instance, if we study a particle of mass m in Euclidean 3-space, the position is given by three numbers x_1, x_2, x_3 (really, coordinates with respect to the standard basis) and the momentum by another three numbers p_1, p_2, p_3 (coordinates with respect to the basis $\{dx_1, dx_2, dx_3\}$ in the cotangent space). See also Example 3.4.20. We will come back to such matters when we have talked about Riemannian metrics.

5.6.1 The Tautological One-Form

If M is an m-dimensional smooth manifold, T^*M is a $2m$-dimensional smooth manifold. This manifold has an especially important one-form $\theta_M \colon T^*M \to T^*T^*M$, called the *tautological one-form* (or *canonical one-form* or *Liouville one-form* or *symplectic potential* – a dear child has many names). For each point $(p, d\phi) \in T^*M$ in the total space of the cotangent bundle we define an element in $T^*_{(p,d\phi)}T^*M$ as follows: consider the map $T\pi \colon T(T^*M) \to TM$ induced by the projection $\pi \colon T^*M \to M$. By the isomorphism $\alpha_p(M) \colon T^*_pM \cong (T_pM)^*$, the cotangent vector $d\phi$ corresponds to the linear map $T_pM \to \mathbf{R}$ sending $[\gamma]$ to $(\phi\gamma)'(0)$. By composing these maps

$$T_{(p,d\phi)}T^*M \to T_pM \to \mathbf{R}$$

we have an element $\theta_M(p, d\phi) \in T^*_{(p,d\phi)}T^*M \cong (T_{(p,d\phi)}T^*M)^*$ (the isomorphism is the inverse of $\alpha_{T^*_pM,(p,d\phi)}$).

Exercise 5.6.6 Show that the procedure above gives a one-form θ_M on T^*M (that is a smooth section of the projection $T^*(T^*M) \to T^*M$).

6 Constructions on Vector Bundles

A good way to think of a vector bundle is as a family of vector spaces indexed over a base space. All constructions we wish to perform on the individual vector spaces should conform with the indexation in that they should vary continuously or smoothly from point to point. This means that, in all essence, the "natural" (in a precise sense) constructions we know from linear algebra have their counterparts for bundles – allowing us to "calculate" with vector bundles (we can "add" and "multiply" them, plus perform a lot of other operations). The resulting theory gives deep information about the base space, as well as allowing us to construct some important mathematical objects. We start this study in this chapter.

6.1 Subbundles and Restrictions

There is a variety of important constructions we need to address. The first of these has been lying underneath the surface for some time.

Definition 6.1.1 Let $\pi : E \to X$ be a rank-n vector bundle. A rank-k (or k-dimensional) *subbundle* of this vector bundle is a subset $E' \subseteq E$ such that around any point in X there is a bundle chart (h, U) for E such that

$$h(\pi^{-1}(U) \cap E') = U \times (\mathbf{R}^k \times \{0\}) \subseteq U \times \mathbf{R}^n.$$

See Figure 6.1.

Note 6.1.2 It makes sense to call such a subset $E' \subseteq E$ a subbundle, since we see that the bundle charts, restricted to E', define a vector bundle structure on $\pi|_{E'} : E' \to X$ which is smooth if we start out with a smooth atlas.

Example 6.1.3 Consider the trivial bundle $S^1 \times \mathbf{C} \to S^1$. The tautological line bundle $\eta_1 \to \mathbf{RP}^1 \cong S^1$ of Example 5.1.3 can be thought of as the subbundle given by

$$\{(e^{i\theta}, te^{i\theta/2}) \in S^1 \times \mathbf{C} \mid t \in \mathbf{R}\} \subseteq S^1 \times \mathbf{C}.$$

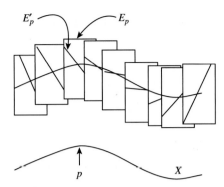

Figure 6.1. A rank-1 subbundle in a rank-2 vector bundle: pick out a one-dimensional linear subspace of every fiber in a continuous manner.

Exercise 6.1.4   Spell out the details of the previous example. Also show that

$$\eta_n = \left\{ ([p], \lambda p) \in \mathbf{RP}^n \times \mathbf{R}^{n+1} \mid p \in S^n, \lambda \in \mathbf{R} \right\} \subseteq \mathbf{RP}^n \times \mathbf{R}^{n+1}$$

is a subbundle of the trivial bundle $\mathbf{RP}^n \times \mathbf{R}^{n+1} \to \mathbf{RP}^n$. Don't look at it before you have done the exercise, but for later reference explicit charts are given in the hint.

Definition 6.1.5   Given a bundle $\pi \colon E \to X$ and a subspace $A \subseteq X$, the *restriction* to A is the bundle

$$\pi_A \colon E_A \to A,$$

where $E_A = \pi^{-1}(A)$ and $\pi_A = \pi|_{\pi^{-1}(A)}$ (Figure 6.2).

As before, in the special case where A is a single point $p \in X$ we write $E_p = \pi^{-1}(p)$ (instead of $E_{\{p\}}$). Occasionally it is typographically convenient to write $E|_A$ instead of E_A.

Note 6.1.6   We see that the restriction is a new vector bundle, and the inclusion

$$
\begin{array}{ccc}
E_A & \overset{\subseteq}{\longrightarrow} & E \\
{\scriptstyle \pi_A}\downarrow & & \downarrow{\scriptstyle \pi} \\
A & \overset{\subseteq}{\longrightarrow} & X
\end{array}
$$

is a bundle morphism inducing an isomorphism (the identity) on every fiber.

Example 6.1.7   Let $N \subseteq M$ be a smooth submanifold. Then we can restrict the tangent bundle on M to N and get

$$(TM)|_N \to N.$$

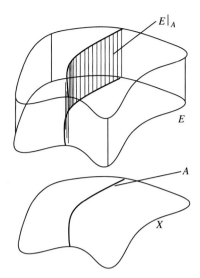

Figure 6.2. The restriction of a bundle $E \to X$ to a subset $A \subseteq X$.

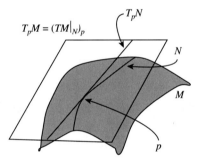

Figure 6.3. In a submanifold $N \subseteq M$ the tangent bundle of N is naturally a subbundle of the tangent bundle of M restricted to N.

We see that $TN \subseteq TM|_N$ is a smooth subbundle (Figure 6.3).

Definition 6.1.8 A bundle morphism

$$
\begin{array}{ccc}
E_1 & \xrightarrow{\ f\ } & E_2 \\
{\scriptstyle \pi_1}\downarrow & & \downarrow{\scriptstyle \pi_2} \\
X_1 & \longrightarrow & X_2
\end{array}
$$

is said to be of *constant rank r* if, restricted to each fiber, f is a linear map of rank r.

Note that this is a generalization of our concept of constant rank of smooth maps.

Theorem 6.1.9 (Rank Theorem for Bundles)　*Consider a bundle morphism*

$$E_1 \xrightarrow{\quad f \quad} E_2$$
$$\pi_1 \searrow \quad \swarrow \pi_2$$
$$X$$

over a space X with constant rank r. Then around any point $p \in X$ there are bundle charts (h, U) and (g, U) such that

$$E_1|_U \xrightarrow{\quad f|_U \quad} E_2|_U$$

$$h \downarrow \qquad\qquad\qquad\qquad\qquad\qquad g \downarrow$$

$$U \times \mathbf{R}^m \xrightarrow{(u,(t_1,\dots,t_m)) \mapsto (u,(t_1,\dots,t_r,0,\dots,0))} U \times \mathbf{R}^n$$

commutes. If we are in a smooth situation, these bundle charts may be chosen to be smooth.

Proof.　This is a local question, so translating via arbitrary bundle charts we may assume that we are in the trivial situation

$$U' \times \mathbf{R}^m \xrightarrow{\quad f \quad} U' \times \mathbf{R}^n$$
$$\mathrm{pr}_{U'} \searrow \quad \swarrow \mathrm{pr}_{U'}$$
$$U'$$

with $f(u, v) = (u, (f_u^1(v), \dots, f_u^n(v)))$, and rk $f_u = r$. By a choice of bases on $\mathbf{R}^m$ and $\mathbf{R}^n$ we may assume that f_u is represented by a matrix

$$\begin{bmatrix} A(u) & B(u) \\ C(u) & D(u) \end{bmatrix}$$

with $A(p) \in \mathrm{GL}_r(\mathbf{R})$ and $D(p) = C(p)A(p)^{-1}B(p)$ (the last equation follows as the rank rk f_p is r). We change the bases so that this is actually true in the standard bases.

Let $p \in U \subseteq U'$ be the open set $U = \{u \in U' \mid \det(A(u)) \neq 0\}$. Then again $D(u) = C(u)A(u)^{-1}B(u)$ on U. Let

$$h \colon U \times \mathbf{R}^m \to U \times \mathbf{R}^m, \qquad h(u, v) = (u, h_u(v))$$

be the homeomorphism where h_u is given by the matrix

$$\begin{bmatrix} A(u) & B(u) \\ 0 & I \end{bmatrix}.$$

Let

$$g \colon U \times \mathbf{R}^n \to U \times \mathbf{R}^n, \qquad g(u, w) = (u, g_u(w))$$

be the homeomorphism where g_u is given by the matrix

$$\begin{bmatrix} I & 0 \\ -C(u)A(u)^{-1} & I \end{bmatrix}.$$

Then $gfh^{-1}(u, v) = (u, (gfh^{-1})_u(v))$, where $(gfh^{-1})_u$ is given by the matrix

$$
\begin{aligned}
&\begin{bmatrix} I & 0 \\ -C(u)A(u)^{-1} & I \end{bmatrix} \begin{bmatrix} A(u) & B(u) \\ C(u) & D(u) \end{bmatrix} \begin{bmatrix} A(u) & B(u) \\ 0 & I \end{bmatrix}^{-1} \\
&= \begin{bmatrix} I & 0 \\ -C(u)A(u)^{-1} & I \end{bmatrix} \begin{bmatrix} A(u) & B(u) \\ C(u) & D(u) \end{bmatrix} \begin{bmatrix} A(u)^{-1} & -A(u)^{-1}B(u) \\ 0 & I \end{bmatrix} \\
&= \begin{bmatrix} I & 0 \\ -C(u)A(u)^{-1} & I \end{bmatrix} \begin{bmatrix} I & 0 \\ C(u)A(u)^{-1} & 0 \end{bmatrix} \\
&= \begin{bmatrix} I & 0 \\ 0 & 0 \end{bmatrix}
\end{aligned}
$$

as claimed (the right-hand lower zero in the answer is really a $0 = -C(u)A(u)^{-1}B(u) + D(u)$). $\blacksquare$

Recall that, if $f: V \to W$ is a linear map of vector spaces, then the *kernel* (or *null space*) is the subspace

$$\ker\{f\} = \{v \in V \mid f(v) = 0\} \subseteq V$$

and the *image* (or *range*) is the subspace

$$\text{Im}\{f\} = \{w \in W \mid \text{there is a } v \in V \text{ such that } w = f(v)\}.$$

We will frequently use the fact that (if V is finite-dimensional)

$$\dim \ker\{f\} - \dim V + \dim \text{Im}\{f\} = 0.$$

Corollary 6.1.10   *If*

is a bundle morphism of constant rank, then the kernel

$$\bigcup_{p \in X} \ker\{f_p\} \subseteq E_1$$

and image

$$\bigcup_{p \in X} \text{Im}\{f_p\} \subseteq E_2$$

are subbundles.

Exercise 6.1.11

Let $\pi \colon E \to X$ be a vector bundle over a connected space X. Assume given a bundle morphism

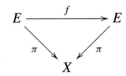

with $f \circ f = f$ (f is "idempotent"). Prove that f has constant rank.

Exercise 6.1.12

Let $\pi \colon E \to X$ be a vector bundle over a connected space X. Assume given a bundle morphism

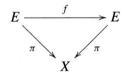

with $f \circ f = \mathrm{id}_E$. Prove that the space of fixed points

$$E^{\{f\}} = \{e \in E \mid f(e) = e\}$$

is a subbundle of E.

Exercise 6.1.13

Show that $f \colon T\mathbf{R} \to T\mathbf{R}$ given by $f([\gamma]) = [t \mapsto \gamma(\gamma(0) \cdot t)]$ is a well-defined bundle morphism, but that f does not have constant rank and neither the kernel nor the image of f is a subbundle.

Exercise 6.1.14

Let $f \colon E \to M$ be a smooth vector bundle of rank k. Show that the *vertical bundle*

$$V = \{v \in TE \mid Tf(v) = 0\} \subseteq TE$$

is a smooth subbundle of $TE \to E$.

6.2 The Induced Bundle

Definition 6.2.1 Assume given a bundle $\pi \colon E \to Y$ and a continuous map $f \colon X \to Y$. Let the *fiber product* of f and π be the space

$$f^*E = X \times_Y E = \{(x, e) \in X \times E \mid f(x) = \pi(e)\}$$

(topologized as a subspace of $X \times E$), and let *the induced bundle* be the projection

$$f^*\pi \colon f^*E \to X, \qquad (x, e) \mapsto x.$$

Note 6.2.2 Note that the fiber over $x \in X$ may be identified with the fiber over $f(x) \in Y$.

The reader may recognize the fiber product $X \times_Y E$ from Exercise 4.7.11, where we showed that if the contributing spaces are smooth then the fiber product is often smooth too.

Lemma 6.2.3 *If $\pi : E \to Y$ is a vector bundle and $f : X \to Y$ a continuous map, then*

$$f^*\pi : f^*E \to X$$

is a vector bundle and the projection $f^*E \to E$ defines a bundle morphism

$$
\begin{array}{ccc}
f^*E & \longrightarrow & E \\
{\scriptstyle f^*\pi} \downarrow & & {\scriptstyle \pi} \downarrow \\
X & \xrightarrow{\;f\;} & Y
\end{array}
$$

inducing an isomorphism on fibers. If the input is smooth the output is smooth too.

Proof. Let $p \in X$ and let (h, V) be a bundle chart

$$h : \pi^{-1}(V) \to V \times \mathbf{R}^k$$

such that $f(p) \in V$. Then $U = f^{-1}(V)$ is an open neighborhood of p. Note that

$$
\begin{aligned}
(f^*\pi)^{-1}(U) &= \{(u, e) \in X \times E \mid f(u) = \pi(e) \in V\} \\
&= \{(u, e) \in U \times \pi^{-1}(V) \mid f(u) = \pi(e)\} \\
&= U \times_V \pi^{-1}(V)
\end{aligned}
$$

and we define

$$
f^*h : (f^*\pi)^{-1}(U) = U \times_V \pi^{-1}(V) \to U \times_V (V \times \mathbf{R}^k) \cong U \times \mathbf{R}^k,
$$
$$
(u, e) \mapsto (u, h(e)) \leftrightarrow (u, h_{\pi(e)}e).
$$

Since h is a homeomorphism f^*h is a homeomorphism (smooth if h is), and since $h_{\pi(e)}$ is an isomorphism (f^*h) is an isomorphism on each fiber. The rest of the lemma now follows automatically. ∎

Theorem 6.2.4 *Let*

$$
\begin{array}{ccc}
E' & \xrightarrow{\;\tilde{f}\;} & E \\
{\scriptstyle \pi'} \downarrow & & {\scriptstyle \pi} \downarrow \\
X' & \xrightarrow{\;f\;} & X
\end{array}
$$

be a bundle morphism. Then there is a factorization

$$
\begin{array}{ccccc}
E' & \longrightarrow & f^*E & \longrightarrow & E \\
{\scriptstyle \pi'} \downarrow & & {\scriptstyle f^*\pi} \downarrow & & {\scriptstyle \pi} \downarrow \\
X' & =\!\!=\!\!= & X' & \xrightarrow{\;f\;} & X.
\end{array}
$$

Proof. Let

$$E' \to X' \times_X E = f^*E,$$
$$e \mapsto (\pi'(e), \tilde{f}(e)).$$

This is well defined since $f(\pi'(e)) = \pi(\tilde{f}(e))$. It is linear on the fibers since the composition

$$(\pi')^{-1}(p) \to (f^*\pi)^{-1}(p) \cong \pi^{-1}(f(p))$$

is nothing but $\tilde{f}_p$.   ∎

Exercise 6.2.5   Let $i : A \subseteq X$ be an inclusion and $\pi : E \to X$ a vector bundle. Prove that the induced and the restricted bundles over A are isomorphic, $i^*E \cong E_A$.

Exercise 6.2.6   Show the following statement: if

$$\begin{array}{ccccc}
E' & \xrightarrow{h} & \tilde{E} & \xrightarrow{g} & E \\
\pi' \downarrow & & \tilde{\pi} \downarrow & & \pi \downarrow \\
X' & = & X' & \xrightarrow{f} & X
\end{array}$$

is a factorization of $(f, \tilde{f})$, then there is a unique bundle map

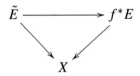

such that

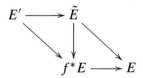

commutes.
   As a matter of fact, you could characterize the induced bundle by this property.

Exercise 6.2.7   Show that, if $f : Y \to X$ is a map and $g : E \to F$ represents a bundle morphism over X, then you have a bundle morphism $f^*g : f^*E \to f^*F$ over Y (with a slight conflict of notation) in a manner that takes composites to composites and identity to identity and so that

$$\begin{array}{ccc}
f^*E & \xrightarrow{f^*g} & f^*F \\
\downarrow & & \downarrow \\
E & \xrightarrow{g} & F
\end{array}$$

commutes, where the vertical maps are the projections. If g is an isomorphism, show that so is f^*g.

Exercise 6.2.8 Show that, if $E \to X$ is a trivial vector bundle and $f : Y \to X$ a map, then $f^*E \to Y$ is trivial.

Exercise 6.2.9 Let $E \to Z$ be a vector bundle and let

$$X \xrightarrow{f} Y \xrightarrow{g} Z$$

be maps. Show that $((gf)^*E \to X) \cong (f^*(g^*E) \to X)$.

Exercise 6.2.10 Let $\pi : E \to X$ be a vector bundle, $\sigma_0 : X \to E$ the zero section and

$$\pi_0 : E \setminus \sigma_0(X) \to X$$

the restriction of π. Construct a nonvanishing section on $\pi_0^*E \to E \setminus \sigma_0(X)$.

6.3 Whitney Sum of Bundles

Natural constructions you can perform on vector spaces can be made to pass to constructions on vector bundles by applying the constructions on each fiber. As an example, we consider the sum $\oplus$. You should check that you believe the constructions, since we plan to be sketchier in future examples.

Definition 6.3.1 If V_1 and V_2 are vector spaces, then $V_1 \oplus V_2 = V_1 \times V_2$ is the vector space of pairs (v_1, v_2) with $v_j \in V_j$. If $f_j : V_j \to W_j$, $j = 1, 2$ are linear maps, then

$$f_1 \oplus f_2 : V_1 \oplus V_2 \to W_1 \oplus W_2$$

is the linear map which sends (v_1, v_2) to $(f_1(v_1), f_2(v_2))$.

Note 6.3.2 Trivially, $\mathrm{id}_{V_1} \oplus \mathrm{id}_{V_2} = \mathrm{id}_{V_1 \oplus V_2}$ and if $g_i : U_i \to V_i$ and $f_j : V_j \to W_j$, $i = 1, 2$ are linear maps then $f_1 g_1 \oplus f_2 g_2 = (f_1 \oplus g_1)(f_2 \oplus g_2)$.

Note that not all linear maps $V_1 \oplus V_2 \to W_1 \oplus W_2$ are of the form $f_1 \oplus f_2$. For instance, if $V_1 = V_2 = W_1 = W_2 = \mathbf{R}$, then the set of linear maps $\mathbf{R} \oplus \mathbf{R} \to \mathbf{R} \oplus \mathbf{R}$ may be identified (by choosing the standard basis) with the set of 2×2 matrices, whereas the maps of the form $f_1 \oplus f_2$ correspond to the diagonal matrices.

Definition 6.3.3 Let $(\pi_1 : E_1 \to X, \mathcal{A}_1)$ and $(\pi_2 : E_2 \to X, \mathcal{A}_2)$ be vector bundles over a common space X. Let

$$E_1 \oplus E_2 = \coprod_{p \in X} E_{1p} \oplus E_{2p}$$

and let $\pi_1 \oplus \pi_2 : E_1 \oplus E_2 \to X$ send all points in the pth summand to $p \in X$. If $(h_1, U_1) \in \mathcal{A}_1$ and $(h_2, U_2) \in \mathcal{A}_2$ then $(h_1 \oplus h_2, U)$ with $U = U_1 \cap U_2$ is defined by the composite

$$(\pi_1 \oplus \pi_2)^{-1}(U) \qquad\qquad U \times (\mathbf{R}^{n_1} \oplus \mathbf{R}^{n_2})$$
$$\|\qquad\qquad\qquad\qquad\qquad\qquad\quad\|$$
$$\coprod_{p\in U} E_{1p} \oplus E_{2p} \xrightarrow{\;\;\coprod h_{1p}\oplus h_{2p}\;\;} \coprod_{p\in U} \mathbf{R}^{n_1} \oplus \mathbf{R}^{n_2}.$$

This defines a pre-vector bundle, and the associated vector bundle is called the *Whitney sum* of the two vector bundles.

If

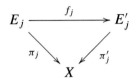

are bundle morphisms over X, then

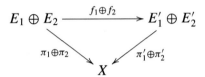

is the bundle morphism defined as $f_1 \oplus f_2$ on each fiber.

Exercise 6.3.4 Check that, if all bundles and morphisms are smooth, then the Whitney sum is a smooth bundle, and that $f_1 \oplus f_2$ is a smooth bundle morphism over X.

Note 6.3.5 Although $\oplus = \times$ for vector spaces, we must not mix them for vector bundles, since $\times$ is reserved for another construction: the product of two bundles $E_1 \times E_2 \to X_1 \times X_2$.

As a matter of fact, the total space $E_1 \oplus E_2$ is the fiber product $E_1 \times_X E_2$.

Exercise 6.3.6 Let
$$\epsilon = \{(p, \lambda p) \in \mathbf{R}^{n+1} \times \mathbf{R}^{n+1} \mid |p| = 1, \lambda \in \mathbf{R}\}.$$

Show that the projection down to S^n defines a trivial bundle.

Definition 6.3.7 A bundle $E \to X$ is called *stably trivial* if there is a trivial bundle $\epsilon \to X$ such that $E \oplus \epsilon \to X$ is trivial.

Exercise 6.3.8 Show that the tangent bundle of the sphere $TS^n \to S^n$ is stably trivial (this is provocative, since even though the tangent bundle of S^2 is nontrivial, we can get a trivial bundle by adding a trivial bundle).

Exercise 6.3.9 Show that the sum of two trivial bundles is trivial. Also show that the sum of two stably trivial bundles is stably trivial.

Exercise 6.3.10

You are given three bundles $\pi_i \colon E_i \to X$, $i = 1, 2, 3$. Show that the set of pairs (f_1, f_2) of bundle morphisms

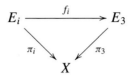

$(i = 1, 2)$ is in one-to-one correspondence with the set of bundle morphisms

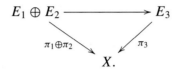

6.4 Linear Algebra on Bundles

There are many constructions on vector spaces that pass on to bundles. We list a few. The examples 1–4, 8 and 9 in Section 6.4.1 will be used in the text, and the others are listed for reference, and for use in exercises.

6.4.1 Constructions on Vector Spaces

1. *The (Whitney) sum.* If V_1 and V_2 are vector spaces, then $V_1 \oplus V_2$ is the vector space of pairs (v_1, v_2) with $v_j \in V_j$. If $f_j \colon V_j \to W_j$ is a linear map $j = 1, 2$, then

$$f_1 \oplus f_2 \colon V_1 \oplus V_2 \to W_1 \oplus W_2$$

is the linear map which sends (v_1, v_2) to $(f_1(v_1), f_2(v_2))$.

2. *The quotient.* If $W \subseteq V$ is a linear subspace we may define the quotient V/W as the set of equivalence classes $V/\sim$ under the equivalence relation that $v \sim v'$ if there is a $w \in W$ such that $v = v' + w$. The equivalence class containing $v \in V$ is written $\bar{v}$. We note that V/W is a vector space with

$$a\bar{v} + b\bar{v}' = \overline{av + bv'}.$$

If $f \colon V \to V'$ is a linear map with $f(W) \subseteq W'$ then f defines a linear map

$$\bar{f} \colon V/W \to V'/W'$$

via the formula $\bar{f}(\bar{v}) = \overline{f(v)}$ (check that this makes sense).

3. *The hom-space.* Let V and W be vector spaces, and let

$$\mathrm{Hom}(V, W)$$

be the set of linear maps $f\colon V \to W$. This is a vector space via the formula $(af + bg)(v) = af(v) + bg(v)$. Note that

$$\mathrm{Hom}(\mathbf{R}^m, \mathbf{R}^n) \cong M_{n\times m}(\mathbf{R}).$$

Also, if $R\colon V \to V'$ and $S\colon W \to W'$ are linear maps, then we get a linear map

$$\mathrm{Hom}(V', W) \xrightarrow{\ \mathrm{Hom}(R,S)\ } \mathrm{Hom}(V, W')$$

by sending $f\colon V' \to W$ to

$$V \xrightarrow{\ R\ } V' \xrightarrow{\ f\ } W \xrightarrow{\ S\ } W'$$

(note that the direction of R is turned around!).

4. *The dual space.* This is a special case of the example above (and was discussed thoroughly in the section following Definition 3.4.10): if V is a vector space, then the dual space is the vector space

$$V^* = \mathrm{Hom}(V, \mathbf{R}).$$

5. *The tensor product.* Let V and W be vector spaces. Consider the set of bilinear maps from $V \times W$ to some other vector space V'. The *tensor product*

$$V \otimes W$$

is the vector space codifying this situation in the sense that giving a bilinear map $V \times W \to V'$ is the same as giving a linear map $V \otimes W \to V'$. With this motivation it is possible to write down explicitly what $V \otimes W$ is: as a set it is the set of all finite linear combinations of symbols $v \otimes w$, where $v \in V$ and $w \in W$ subject to the relations

$$a(v \otimes w) = (av) \otimes w = v \otimes (aw),$$
$$(v_1 + v_2) \otimes w = v_1 \otimes w + v_2 \otimes w,$$
$$v \otimes (w_1 + w_2) = v \otimes w_1 + v \otimes w_2,$$

where $a \in \mathbf{R}$, $v, v_1, v_2 \in V$ and $w, w_1, w_2 \in W$. This is a vector space in the obvious manner, and given linear maps $f\colon V \to V'$ and $g\colon W \to W'$ we get a linear map

$$f \otimes g\colon V \otimes W \to V' \otimes W'$$

by sending $\sum_{i=1}^k v_i \otimes w_i$ to $\sum_{i=1}^k f(v_i) \otimes g(w_i)$ (check that this makes sense!). Note that

$$\mathbf{R}^m \otimes \mathbf{R}^n \cong \mathbf{R}^{mn}$$

and that there are isomorphisms

$$\mathrm{Hom}(V \otimes W, V') \cong \{\text{bilinear maps } V \times W \to V'\}:$$

the bilinear map associated with a linear map $f: V \otimes W \to V'$ sends $(v, w) \in V \times W$ to $f(v \otimes w)$. The linear map associated with a bilinear map $g: V \times W \to V'$ sends $\sum v_i \otimes w_i \in V \otimes W$ to $\sum g(v_i, w_i)$.

For some reason, people tend to try to make tensors into frightening beasts. It doesn't help that the classical literature calls an element of $V^{\otimes n} = V \otimes \cdots \otimes V$ (n tensor factors) a "contravariant n-tensor" and obscures it by double dualizing and considering $((V^*)^{\otimes n})^*$ instead (they are isomorphic when V is finite-dimensional). The vector space of "tensors of type $\binom{k}{l}$" is (in the finite-dimensional case naturally isomorphic to) $\mathrm{Hom}(V^{\otimes k}, V^{\otimes l})$.

6. *The exterior power.* Let V be a vector space. The kth exterior power $\Lambda^k V$ is defined as the quotient of the k-fold tensor product $V^{\otimes k}$ by the subspace generated by the elements $v_1 \otimes v_2 \otimes \cdots \otimes v_k$, where $v_i = v_j$ for some $i \neq j$. The image of $v_1 \otimes v_2 \otimes \cdots \otimes v_k$ in $\Lambda^k V$ is written $v_1 \wedge v_2 \wedge \cdots \wedge v_k$. Note that it follows that $v_1 \wedge v_2 = -v_2 \wedge v_1$ since

$$0 = (v_1 + v_2) \wedge (v_1 + v_2) = v_1 \wedge v_1 + v_1 \wedge v_2 + v_2 \wedge v_1 + v_2 \wedge v_2 = v_1 \wedge v_2 + v_2 \wedge v_1$$

and similarly for more $\wedge$ factors: swapping two entries changes sign.

Note that the dimension of $\Lambda^k \mathbf{R}^n$ is $\binom{n}{k}$. There is a particularly nice isomorphism $\Lambda^n \mathbf{R}^n \to \mathbf{R}$ given by the determinant function.

7. *The symmetric power.* Let V be a vector space. The kth symmetric power $S^k V$ is defined as the quotient of the k-fold tensor product $V^{\otimes k}$ by the subspace generated by the elements $v_1 \otimes v_2 \otimes \cdots \otimes v_i \otimes \cdots \otimes v_j \otimes \cdots \otimes v_k - v_1 \otimes v_2 \otimes \cdots \otimes v_j \otimes \cdots \otimes v_i \otimes \cdots \otimes v_k$.

8. *Alternating forms.* The space of alternating forms $\mathrm{Alt}^k(V)$ on a vector space V is defined to be $(\Lambda^k V)^*$, the dual of the exterior power $\Lambda^k V$ in Section 6.4.1(6). That is, $\mathrm{Alt}^k(V)$ consists of the multilinear maps

$$f: V \times \cdots \times V \to \mathbf{R}$$

(in k V-variables) which are zero on inputs with repeated coordinates.

The space of alternating forms on the tangent space is the natural home of the symbols like $dx\, dy\, dz$ you'll find in elementary multivariable analysis.

Again, the dimension of $\mathrm{Alt}^k(V)$ is $\binom{\dim V}{k}$, and the determinant is a basis element for the one-dimensional vector space $\mathrm{Alt}^n(\mathbf{R}^n)$.

9. *Symmetric bilinear forms.* Let V be a vector space. The space of $SB(V)$ symmetric bilinear forms is the space of bilinear maps $f: V \times V \to \mathbf{R}$ such that $f(v, w) = f(w, v)$. In other words, the space of symmetric bilinear forms is $SB(V) = (S^2 V)^*$.

6.4.2 Constructions on Vector Bundles

When translating these constructions to vector bundles, it is important not only to bear in mind what they do on each individual vector space but also what they do on linear maps. Note that some of the examples "turn the arrows around". The Hom-space in Section 6.4.1(3) is a particular example of this: it "turns the arrows around" in the first variable, but not in the second.

Instead of giving the general procedure for translating such constructions to bundles in general, we do it on the Hom-space, which exhibits all the potential difficult points.

Example 6.4.3 Let $(\pi: E \to X, \mathcal{B})$ and $(\pi': E' \to X, \mathcal{B}')$ be vector bundles of rank m and n. We define a pre-vector bundle

$$\mathrm{Hom}(E, E') = \coprod_{p \in X} \mathrm{Hom}(E_p, E'_p) \to X$$

of rank mn as follows. The projection sends the pth summand to p, and given bundle charts $(h, U) \in \mathcal{B}$ and $(h', U') \in \mathcal{B}'$ we define a bundle chart $(\mathrm{Hom}(h^{-1}, h'), U \cap U')$. On the fiber above $p \in X$,

$$\mathrm{Hom}(h^{-1}, h')_p: \mathrm{Hom}(E_p, E'_p) \to \mathrm{Hom}(\mathbf{R}^m, \mathbf{R}^n) \cong \mathbf{R}^{mn}$$

is given by sending $f: E_p \to E'_p$ to

$$
\begin{array}{ccc}
\mathbf{R}^m & & \mathbf{R}^n \\
{\scriptstyle h_p^{-1}}\big\downarrow & & \big\uparrow{\scriptstyle h'_p} \\
E_p & \xrightarrow{\ f\ } & E'_p.
\end{array}
$$

If $(g, V) \in \mathcal{B}$ and $(g', V') \in \mathcal{B}'$ are two other bundle charts, the transition function becomes

$$p \mapsto \mathrm{Hom}(g_p^{-1}, g'_p)\left(\mathrm{Hom}(h_p^{-1}, h'_p)\right)^{-1} = \mathrm{Hom}(h_p g_p^{-1}, g'_p(h'_p)^{-1}),$$

sending $f: \mathbf{R}^m \to \mathbf{R}^n$ to

$$
\begin{array}{ccc}
\mathbf{R}^m & & \mathbf{R}^n \\
{\scriptstyle g_p^{-1}}\big\downarrow & & \big\uparrow{\scriptstyle g'_p} \\
E_p & & E'_p \\
{\scriptstyle h_p}\big\downarrow & & \big\uparrow{\scriptstyle (h'_p)^{-1}} \\
\mathbf{R}^m & \xrightarrow{\ f\ } & \mathbf{R}^n.
\end{array}
$$

That is, if $W = U \cap U' \cap V \cap V'$, then the transition function

$$W \longrightarrow \mathrm{GL}(\mathrm{Hom}(\mathbf{R}^m, \mathbf{R}^n)) \cong \mathrm{GL}_{mn}(\mathbf{R})$$

is the composite of

(1) the diagonal $W \to W \times W$ sending p to (p, p),
(2) the product of the transition functions

$$W \times W \to \mathrm{GL}(\mathbf{R}^m) \times \mathrm{GL}(\mathbf{R}^n),$$

sending (p, q) to $(g_p h_p^{-1}, h'_q(g'_q)^{-1})$, and

(3) the map

$$GL(\mathbf{R}^m) \times GL(\mathbf{R}^n) \to GL(\text{Hom}(\mathbf{R}^m, \mathbf{R}^n)),$$

sending (A, B) to $\text{Hom}(A, B)$.

The first two are continuous or smooth depending on whether the bundles are topological or smooth. The last map, $GL(\mathbf{R}^m) \times GL(\mathbf{R}^n) \to GL(\text{Hom}(\mathbf{R}^m, \mathbf{R}^n))$, is smooth ($C \mapsto BCA$ is a linear transformation on $\text{Hom}(\mathbf{R}^m, \mathbf{R}^n)$ which depends smoothly on A and B since the algebraic operations are smooth).

In effect, the transition functions of $\text{Hom}(E, E') \to X$ are smooth (resp. continuous) if the transition functions of $E \to X$ and $E' \to X$ are smooth (resp. continuous).

It is worth pausing a bit at this point. The three-point approach above will serve us in many cases, so you should review it carefully. In particular, the smoothness of the third map and the equation $\text{Hom}(g_p^{-1}, g_p')\left(\text{Hom}(h_p^{-1}, h_p')\right)^{-1} = \text{Hom}(h_p g_p^{-1}, g_p'(h_p')^{-1})$ are pure linear algebra features of the Hom-construction that have nothing to do with the vector bundles.

Exercise 6.4.4

Let $E \to X$ and $E' \to X$ be vector bundles. Show that there is a one-to-one correspondence between bundle morphisms

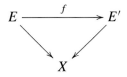

and sections of $\text{Hom}(E, E') \to X$.

Exercise 6.4.5

Convince yourself that the construction of $\text{Hom}(E, E') \to X$ outlined above really gives a vector bundle, and that if

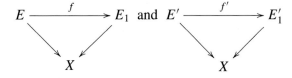

are bundle morphisms, we get another

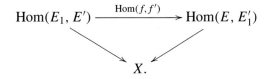

Exercise 6.4.6

Let $L \to X$ be a line bundle. Show that the associated Hom-bundle $\mathrm{Hom}(L, L) \to X$ is trivial both in the continuous situation and in the smooth situation.

Exercise 6.4.7

Write out the definition of the *quotient bundle*, and show that, if

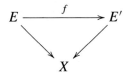

is a bundle map, and $F \subseteq E$ and $F' \subseteq E'$ are subbundles such that $\mathrm{Im}\{f|_F\} \subseteq F'$, then we get a bundle morphism

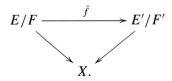

Example 6.4.8   Given a bundle $E \to X$, the *dual bundle* $E^* \to X$ is important in many situations. If (h, U) is a bundle chart, then we get a bundle chart for the dual bundle

$$(E^*)_U = \coprod_{p \in U} E_p^* \xrightarrow{\coprod (h_p^{-1})^*} \coprod_{p \in U} (\mathbf{R}^k)^* = U \times (\mathbf{R}^k)^*$$

(choose a fixed isomorphism $(\mathbf{R}^k)^* \cong \mathbf{R}^k$).

Exercise 6.4.9

Check that the bundle charts proposed for the dual bundle actually give a bundle atlas, and that this atlas is smooth if the original bundle was smooth.

Exercise 6.4.10

For those who read the section on the cotangent bundle $T^*M \to M$ associated with a smooth n-manifold M: prove that the maps of Proposition 3.4.14

$$\alpha_p \colon T_p^*M \to (T_pM)^*, \quad d\phi \mapsto \{[\gamma] \mapsto (\phi x)'(0)\}$$

induce an isomorphism from the cotangent bundle to the dual of the tangent bundle.

Given Exercise 6.4.10, readers who have *not* studied the cotangent bundle are free to define it in the future as the dual of the tangent bundle. Recall that the elements of the cotangent bundle are called 1-*forms*.

Exercise 6.4.11

Given a bundle $E \to X$, write out the definition of the associated *symmetric bilinear forms bundle* $SB(E) \to X$.

Example 6.4.12 An alternating k-*form* (or just k-form) is an element in $\mathrm{Alt}^k(TM)$ (see Section 6.4.1(8). These are the main objects of study when doing analysis of manifolds (integration, etc.).

Exercise 6.4.13 Write out the definition of the bundle of alternating k-forms, and, if you are still not bored stiff, do some more examples. If you are really industrious, find out on what level of generality these ideas really work, and prove it there.

Exercise 6.4.14 Let $L \to M$ be a line bundle. Show that the tensor product $L \otimes L \to M$ is also a line bundle and that all the transition functions in the associated (non-maximal) bundle atlas on $L \otimes L \to M$ have only positive values.

6.5 Normal Bundles

We will later discuss Riemannian structures and more generally fiber metrics over smooth manifolds. This will give us the opportunity to discuss inner products, and in particular questions pertaining to orthogonality, in the fibers. That such structures exist over smooth manifolds is an artifact of the smooth category, in which local smooth data occasionally can be patched together to give global smooth structures.

However, there is a formulation of these phenomena which does not depend on inner products, but rather uses quotient bundles.

Definition 6.5.1 Let $N \subseteq M$ be a smooth submanifold. The *normal bundle* $\perp N \to N$ is defined as the quotient bundle $(TM|_N)/TN \to N$ (see Exercise 6.4.7 and Figure 6.4).

More generally, if $f: N \to M$ is an imbedding, we define the normal bundle $\perp^f N \to N$ to be the bundle $(f^*TM)/TN \to N$.

It turns out that there is an important consequence of transversality pertaining to normal bundles.

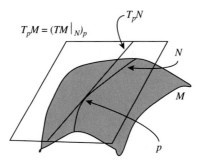

Figure 6.4. In a submanifold $N \subseteq M$ the tangent bundle of N is naturally a subbundle of the tangent bundle of M restricted to N, and the normal bundle is the quotient on each fiber.

Theorem 6.5.2 *Assume* $f: N \to M$ *is transverse to a k-codimensional submanifold* $L \subseteq M$ *and that* $f(N) \cap L \neq \emptyset$. *Then* $f^{-1}(L) \subseteq N$ *is a k-codimensional submanifold and there is an isomorphism*

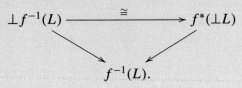

Proof. The first part is simply Theorem 4.5.5. For the statement about normal bundles, remember that $\perp L = (TM|_L)/TL$ and $\perp f^{-1}(L) = (TN|_{f^{-1}(L)})/T(f^{-1}(L))$ and consider the diagram

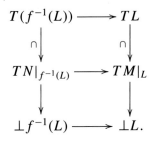

Transversality gives that the map from $TN|_{f^{-1}(L)}$ to $\perp L$ is surjective on every fiber, and so – for dimensional reasons – $\perp f^{-1}(L) \to \perp L$ is an isomorphism on every fiber. This then implies that $\perp f^{-1}(L) \to f^*(\perp L)$ must be an isomorphism by Lemma 5.3.12. ∎

Corollary 6.5.3 *Consider a smooth map* $f: N \to M$ *and a regular value* $q \in M$. *Then the normal bundle* $\perp f^{-1}(q) \to f^{-1}(q)$ *is trivial.*

Note 6.5.4 In particular, this shows that the normal bundle of $S^n \subseteq \mathbf{R}^{n+1}$ is trivial. Also it shows that the normal bundle of $O(n) \subseteq M_n(\mathbf{R})$ is trivial, and so are all the other manifolds we constructed in Chapter 4 as the inverse image of regular values.

In Exercise 6.3.8 we showed that the tangent bundle of S^n is stably trivial, and an analysis of that proof gives an isomorphism between $T\mathbf{R}^{n+1}|_{S^n}$ and $TS^n \oplus \perp S^n$. This "splitting" is a general phenomenon and is a result of the flexibility of the smooth category alluded to at the beginning of this section. We will return to such issues in Section 6.6 when we discuss Riemannian structures.

Exercise 6.5.5

Let M be a smooth manifold, and consider M as a submanifold by imbedding it as the diagonal in $M \times M$ (i.e., as the set $\{(p, p) \in M \times M\}$: show that it is a smooth submanifold). Prove that the normal bundle $\perp M \to M$ is isomorphic to $TM \to M$.

Exercise 6.5.6   Consider the tautological line bundle $\eta_1 \to S^1$. Show that $\eta_1 \oplus \eta_1 \to S^1$ is trivial.

6.6 Riemannian Metrics

In differential geometry one works with more highly structured manifolds than in differential topology. In particular, all manifolds should come equipped with metrics on the tangent spaces which vary smoothly from point to point. This is what is called a Riemannian manifold, and is crucial to many applications.

We will eventually show in Theorem 8.3.1 that all smooth manifolds *have* a structure of a Riemannian manifold. However, there is a huge difference between merely saying that a given manifold has *some* Riemannian structure and actually working with manifolds with a *chosen* Riemannian structure.

Recall from Section 6.4.1(9) that, if V is a vector space, then $SB(V)$ is the vector space of all symmetric bilinear forms $g: V \times V \to \mathbf{R}$, i.e., functions g such that $g(v, w) = g(w, v)$ and which are linear in each variable. In particular, there is a linear isomorphism between $SB(\mathbf{R}^n)$ and the vector space $\mathrm{Sym}(n)$ of symmetric $n \times n$ matrices, given by sending a symmetric bilinear form g to the symmetric matrix $(g(e_i, e_j))$; the inverse sends a symmetric matrix A to the symmetric bilinear form $(v, w) \mapsto \langle v, w \rangle_A = v^\mathsf{T} A w$.

Recall that this lifts to the level of bundles: if $\pi: E \to X$ is a bundle, we get an associated symmetric bilinear forms bundle $SB(\pi): SB(E) \to X$ (see Exercise 6.4.11). A more involved way of saying this is $SB(E) = (S^2 E)^* \to X$ in the language of Section 6.4.1(4) and Section 6.4.1(7).

Definition 6.6.1   Let V be a vector space. An *inner product* is a symmetric bilinear form $g \in SB(V)$ which is *positive definite*, i.e., we have that $g(v, v) \geq 0$ for all $v \in V$ and $g(v, v) = 0$ only if $v = 0$.

Example 6.6.2   If A is a symmetric $n \times n$ matrix, then $\langle v, w \rangle_A = v^\mathsf{T} A w$ defines an inner product $\langle , \rangle_A \in SB(\mathbf{R}^n)$ exactly if A is positive definite (all eigenvalues are positive). In particular, if A is the identity matrix we get the standard inner product on $\mathbf{R}^n$.

Definition 6.6.3   A *fiber metric* on a vector bundle $\pi: E \to X$ is a section $g: X \to SB(E)$ on the associated symmetric bilinear forms bundle, such that for every $p \in X$ the associated symmetric bilinear form $g_p: E_p \times E_p \to \mathbf{R}$ is positive definite. The fiber metric is smooth if both the vector bundle $E \to X$ and the section g are smooth.

A fiber metric is often called a *Riemannian metric*, although many authors reserve this notion for a fiber metric on the tangent bundle of a smooth manifold.

Definition 6.6.4   A *Riemannian manifold* is a smooth manifold with a smooth fiber metric on the tangent bundle.

Example 6.6.5 Let $M = \mathbf{R}^n$ and consider the standard trivialization of the tangent bundle given by $T\mathbf{R}^n \cong \mathbf{R}^n \times \mathbf{R}^n$ sending $[\gamma]$ to $(\gamma(0), \gamma'(0))$. The usual dot product on $\mathbf{R}^n$ gives a fiber metric on the product bundle and hence a Riemannian metric on $\mathbf{R}^n$:

$$g([\gamma_1], [\gamma_2]) = \gamma_1'(0) \cdot \gamma_2'(0).$$

Somewhat more generally, parallelizable manifolds have especially attractive Riemannian metrics: given a global trivialization $TM \cong M \times \mathbf{R}^n$ we choose the usual dot product for $\mathbf{R}^n$ and transport this back. Of course, these considerations carry over to trivial vector bundles in general.

Fiber metrics are used to confuse bundles with their dual.

Lemma 6.6.6 *Let V be a (finite-dimensional) vector space with an inner product $\langle -, - \rangle$. Then the assignment*

$$V \to V^*, \qquad v \mapsto \langle v, - \rangle$$

(where $\langle v, - \rangle \colon V \to \mathbf{R}$ sends w to $\langle v, w \rangle$) is a linear isomorphism.

Proof. The bilinearity of the inner product ensures that the map $V \to V^*$ is linear and well defined. The non-degenerate property of the inner product is equivalent to the injectivity of $V \to V^*$, and, since any injective linear map of vector spaces of equal finite dimension is an isomorphism, $V \to V^*$ is an isomorphism. ∎

Since a fiber metric $p \mapsto g_p$ varies smoothly/continuously in p, this assembles to an isomorphism of bundles:

Corollary 6.6.7 *Let $E \to M$ be a vector bundle and g a fiber metric. Then*

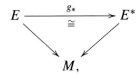

defined by $g_* \colon E_p \cong (E_p)^*$ with $g_*(v) = g_p(v, -)$, is an isomorphism. If (M, g) is a Riemannian manifold this gives an isomorphism between the tangent bundle and its dual

$$TM \xrightarrow[\cong]{g_*} (TM)^*.$$

If you prefer the cotangent bundle to the dual tangent bundle you should follow through with the inverse of the natural isomorphism $\alpha \colon T^*M \to (TM)^*$ of Proposition 3.4.14 (and expanded on in Exercise 6.4.10) sending $d\phi \in T_p^*M$ to the linear map $T_pM \to \mathbf{R}$, $[\gamma] \mapsto (\phi\gamma)'(0)$. Consequently, the Riemannian metric provides us with an isomorphism

$$TM \xrightarrow[\cong]{g_*} (TM)^* \xleftarrow[\cong]{\alpha} T^*M,$$

which is used for all it's worth (and a bit more) in many expositions. The notion of the *gradient* of a real-valued smooth function f is nothing but the differential df carried over with this isomorphism.

Definition 6.6.8 Let (M, g) be a Riemannian manifold and let $f : M \to \mathbf{R}$ be a smooth function. The *gradient of f* is the vector field

$$\operatorname{grad} f = g_*^{-1} \alpha \, df : M \to TM.$$

Another way of saying this is that $\operatorname{grad} f$ is the unique vector field such that if X is any other vector field, then $df(X) = g(\operatorname{grad} f, X)$.

Example 6.6.9 Let $M = \mathbf{R}^n$ equipped with the Riemannian metric of Example 6.6.5 given by the dot product on $T_p\mathbf{R}^n \cong \mathbf{R}^n$:

$$g_p([\gamma_1], [\gamma_2]) = \gamma_1'(0) \cdot \gamma_2'(0).$$

Then the composite isomorphism

$$\mathbf{R}^n \xleftarrow[\cong]{A_{\mathrm{id}}} T_p\mathbf{R}^n \xrightarrow[\cong]{g_*} (T_p\mathbf{R}^n)^* \xleftarrow[\cong]{A_{\mathrm{id}}^*} (\mathbf{R}^n)^* \xrightarrow{\cong} M_{1 \times n}\mathbf{R}$$

is a very complicated way of expressing transposition. If $f : M \to \mathbf{R}$ is smooth, then $d_p f \in T_p^* M$ and $\operatorname{grad}_p f \in T_p M$ both correspond to the Jacobi matrix $D_p f \in M_{1 \times n}\mathbf{R}$.

Example 6.6.10 Consider the function $f : S^1 \times S^1 \to \mathbf{R}$ (whose differential is written out in Example 4.1.11) of the robot example. Since $S^1 \times S^1$ is parallelizable we see that the gradient is given by $S^1 \times S^1 \to T(S^1 \times S^1) \cong (S^1 \times S^1) \times R^2$,

$$(e^{i\theta}, e^{i\phi}) \mapsto \left((e^{i\theta}, e^{i\phi}), \frac{1}{f(e^{i\theta}, e^{i\phi})} \begin{bmatrix} 3\sin\theta - \cos\phi\sin\theta + \sin\phi\cos\theta \\ 3\sin\phi - \cos\theta\sin\phi + \sin\theta\cos\phi \end{bmatrix} \right).$$

Exercise 6.6.11 Make a visual representation of the gradient field of the function $f : S^1 \times S^1 \to \mathbf{R}$ (whose differential is written out in Example 4.1.11) of the robot example. Pay special attention to the behavior near the critical points.

Normal bundles in general were introduced in Section 6.5, but in the presence of a fiber metric things become somewhat less abstract.

Definition 6.6.12 Given a bundle $\pi : E \to X$, a fiber metric g and a subbundle $F \subseteq E$, we define the *normal bundle with respect to g* of $F \subseteq E$ to be the subset

$$F^\perp = \coprod_{p \in X} F_p^\perp$$

given by taking the orthogonal complement of $F_p \subseteq E_p$ (relative to the inner product $g(p)$).

Lemma 6.6.13 *Given a bundle $\pi : E \to X$ with a fiber metric g and a subbundle $F \subseteq E$,*

(1) the normal bundle $F^\perp \subseteq E$ is a subbundle;
(2) the composite

$$F^\perp \subseteq E \to E/F$$

is an isomorphism of bundles over X;
(3) the bundle morphism $F \oplus F^\perp \to E$ induced by the inclusions is an isomorphism over X;
(4) the bundle $E \to X$ has an atlas whose transition functions map to the orthogonal group.

The lemma also holds in the smooth case.

Proof. Choose a bundle chart (h, U) such that

$$h(F|_U) = U \times (\mathbf{R}^k \times \{0\}) \subseteq U \times \mathbf{R}^n.$$

Let $v_j(p) = h^{-1}(p, e_j) \in E_p$ for $p \in U$. Then $(v_1(p), \ldots, v_n(p))$ is a basis for E_p, whereas $(v_1(p), \ldots, v_k(p))$ is a basis for F_p. Perform the Gram–Schmidt process with respect to the metric $g(p)$ to transform these bases to orthogonal bases $(v_1'(p), \ldots, v_n'(p))$ for E_p, $(v_1'(p), \ldots, v_k'(p))$ for F_p and $(v_{k+1}'(p), \ldots, v_n'(p))$ for $F_p^\perp$.

We define a new bundle chart (h', U) by

$$h' : E|_U \to U \times \mathbf{R}^n,$$

$$\sum_{i=1}^n a_i v_i'(p) \mapsto (p, (a_1, \ldots, a_n))$$

$((h', U)$ is a bundle chart since the metric varies continuously/smoothly with p; explicitly, the transition function $p \mapsto h_p h'^{-1}_p$ is given by the upper triangular matrix with ones on the diagonal and for $k < i$ the (k, i) entry is $g(p)(v_i(p), v_k(p))/(g(p)(v_i(p), v_i(p)))$, which restricts to an isomorphism between $F^\perp|_U$ and $U \times (\{0\} \times \mathbf{R}^{n-k})$. Hence, $F^\perp \subseteq E$ is a subbundle.

For the second claim, observe that $\dim F_p^\perp = \dim E_p/F_p$, and so the claim follows if the map $F^\perp \subseteq E \to E/F$ is injective on every fiber, but this is true since $F_p \cap F_p^\perp = \{0\}$.

For the third claim, note that the map in question induces a linear map on every fiber which is an isomorphism, and hence by Lemma 5.3.12 the map is an isomorphism of bundles.

Lastly, if in the Gram–Schmidt process in the proof we also normalize, we have shown that we can always choose an atlas such that the charts (h', U) are orthogonal on every fiber (i.e., $g(p)(e, e') = h'_p(e) \cdot h'_p(e')$) and all the transition functions between maps produced in this fashion would map to the orthogonal group $O(n) \subseteq GL_n(\mathbf{R})$.   ∎

Example 6.6.14   In applications the fiber metric is often given by physical considerations. Consider a particle moving on a manifold M defining a smooth curve $\gamma : \mathbf{R} \to M$. At each point the velocity of the curve defines a tangent vector, and so the curve lifts to a curve on the tangent space $\dot{\gamma} : \mathbf{R} \to TM$ (see Definition 7.1.13 for careful definitions). The dynamics is determined by the energy, and the connection between the metric and the energy is that the norm associated with the metric g at a given point is twice the kinetic energy T, c.f. Example 3.4.20. The "generalized" or "conjugate momentum" in mechanics is then nothing but g_* of the velocity, living in the cotangent bundle T^*M which is often referred to as the "phase space".

For instance, if $M = \mathbf{R}^n$ (with the identity chart) and the mass of the particle is m, the kinetic energy of a particle moving with velocity $v \in T_pM$ at $p \in M$ is $\frac{1}{2}m|v|^2$, and so the appropriate metric is m times the usual Euclidean metric $g_p(v, w) = m \cdot \langle v, w \rangle$ (and in particular independent of p) and the generalized momentum is $m\langle v, - \rangle \in T_p^*M$.

6.7 Orientations[1]

. . . .

The space of alternating forms $\text{Alt}^k(V)$ on a vector space V is defined to be $(\Lambda^k V)^* = \text{Hom}(\Lambda^k V, \mathbf{R})$ (see Section 6.4.1(8)), or, alternatively, $\text{Alt}^k(V)$ consists of the multilinear maps

$$V \times \cdots \times V \to \mathbf{R}$$

in k V-variables which are zero on inputs with repeated coordinates.

In particular, if $V = \mathbf{R}^k$ we have the *determinant function*

$$\det \in \text{Alt}^k(\mathbf{R}^k)$$

given by sending $v_1 \wedge \cdots \wedge v_k$ to the determinant of the $k \times k$ matrix $[v_1 \ldots v_k]$ you get by considering v_i as the ith column.

In fact, $\det : \Lambda^k \mathbf{R}^k \to \mathbf{R}$ is an isomorphism.

Exercise 6.7.1   Check that the determinant actually is an alternating form and an isomorphism.

Definition 6.7.2   An *orientation* on a k-dimensional vector space V is an equivalence class of bases on V, where $(v_1, \ldots, v_k)$ and $(w_1, \ldots, w_k)$ are equivalent if $v_1 \wedge \cdots \wedge v_k = \lambda w_1 \wedge \cdots \wedge w_k$ for some $\lambda > 0$. The equivalence class, or *orientation class*, represented by a basis $(v_1, \ldots, v_k)$ is written $[v_1, \ldots, v_k]$.

. . . .

[1] This section is not used anywhere else and may safely be skipped.

Note 6.7.3 That two bases $(v_1, \ldots, v_k)$ and $(w_1, \ldots, w_k)$ in $\mathbf{R}^k$ define the same orientation class can be formulated by means of the determinant:

$$\det(v_1 \wedge \cdots \wedge v_k)/\det(w_1 \wedge \cdots \wedge w_k) > 0.$$

As a matter of fact, this formula is valid for any k-dimensional vector space V (choose an isomorphism $V \cong \mathbf{R}^k$ and check that the choice turns out not to matter).

Note 6.7.4 On a vector space V there are exactly two orientations. For instance, on $\mathbf{R}^k$ the two orientations are $[e_1, \ldots, e_k]$ and $[-e_1, e_2, \ldots, e_k] = [e_2, e_1, e_3 \ldots, e_k]$.

Note 6.7.5 An *isomorphism* of vector spaces $f \colon V \to W$ sends an orientation $\mathcal{O} = [v_1, \ldots, v_k]$ to the orientation $f\mathcal{O} = [f(v_1), \ldots, f(v_k)]$.

Definition 6.7.6 An *oriented vector space* is a vector space together with a chosen orientation. An isomorphism of oriented vector spaces either *preserves* or *reverses* the orientation.

Definition 6.7.7 Let $E \to X$ be a vector bundle. An *orientation* on $E \to X$ is a family $\mathcal{O} = \{\mathcal{O}_p\}_{p \in X}$ such that $\mathcal{O}_p$ is an orientation on the fiber E_p, and such that around any point $p \in X$ there is a bundle chart (h, U) such that for all $q \in U$ we have that $h_q \mathcal{O}_q = h_p \mathcal{O}_p$.

Definition 6.7.8 A vector bundle is *orientable* if it can be equipped with an orientation.

Example 6.7.9 A trivial bundle is orientable.

Example 6.7.10 Not all bundles are orientable, for instance, the tautological line bundle $\eta_1 \to S^1$ of Example 5.1.3 is not orientable: start choosing orientations, run around the circle, and have a problem.

Definition 6.7.11 A smooth manifold M is *orientable* if the tangent bundle is orientable. An *oriented diffeomorphism* is a diffeomorphism $f \colon M \to N$ such that for all $p \in M$ the tangent map $T_p f$ preserves the orientation.

6.8 The Generalized Gauss Map[2]

The importance of the Grassmann manifolds to bundle theory stems from the fact that in a certain precise sense the bundles over a given manifold M are classified by a set of equivalence classes (called homotopy classes) from M into Grassmann manifolds. This is really cool, but unfortunately beyond the scope of our current

---

[2] This section is not used anywhere else and may safely be skipped.

investigations. We offer a string of exercises as a vague orientation into interesting stuff we can't pursue to the depths it deserves.

Exercise 6.8.1   Recall the Grassmann manifold $\text{Gr}(k, \mathbf{R}^n)$ of all k-dimensional linear subspaces of $\mathbf{R}^n$ defined in Example 2.3.15. Define the *tautological k-plane bundle* over the Grassmann manifold

$$\gamma_n^k \to \text{Gr}(k, \mathbf{R}^n)$$

by setting

$$\gamma_n^k = \{(E, v) \mid E \in \text{Gr}(k, \mathbf{R}^n), v \in E\}.$$

Note that $\gamma_n^1 = \eta_n \to \mathbf{R}\mathrm{P}^n = \text{Gr}(1, \mathbf{R}^{n+1})$. (Hint: use the charts in Example 2.3.15, and let

$$h_V \colon \pi^{-1}(U_V) \to U_V \times V$$

send (E, v) to $(E, \mathrm{pr}_V v)$.)

All smooth manifolds can be imbedded in Euclidean space, so the assumption in the following exercise is not restrictive.

Exercise 6.8.2   Let $M \subseteq \mathbf{R}^{n+k}$ be a smooth n-dimensional submanifold of Euclidean space. Then we define the *generalized Gauss map*

$$
\begin{array}{ccc}
TM & \longrightarrow & \gamma_{n+k}^n \\
\downarrow & & \downarrow \\
M & \longrightarrow & \text{Gr}(n, \mathbf{R}^{n+k})
\end{array}
$$

by sending $p \in M$ to $T_p M \in \text{Gr}(n, \mathbf{R}^{n+k})$ (we consider $T_p M$ as a subspace of $\mathbf{R}^{n+k}$ under the standard identification $T_p \mathbf{R}^{n+k} = \mathbf{R}^{n+k}$) and $[\gamma] \in TM$ to $(T_{\gamma(0)} M, [\gamma])$. Check that it is a bundle morphism and displays the tangent bundle of M as the induced bundle of the tautological n-plane bundle under $M \to \text{Gr}(n, \mathbf{R}^{n+k})$.

More generally, let E be a rank-n subbundle of the trivial bundle $M \times \mathbf{R}^{n+k} \to M$. Define a map $f \colon M \to \text{Gr}(n, \mathbf{R}^{n+k})$ such that $E \to M$ is isomorphic to $f^* \gamma_{n+k}^n \to M$.

So, we get a lot of bundles by pulling back the tautological k-plane bundle over compact manifolds, but there is some repetition. If $f_0, f_1 \colon M \to \text{Gr}(k, \mathbf{R}^n)$ factor through a map $H \colon M \times [0, 1] \to \text{Gr}(k, \mathbf{R}^n)$ (we say that f_0 and f_1 are *homotopic* and H is a *homotopy*) the vector bundles $f_0^* \gamma_n^k$ and $f_1^* \gamma_n^k$ are isomorphic.

The upshot is a classification result: a one-to-one correspondence between isomorphism classes of vector bundles over M and homotopy classes of maps from M to the Grassmannian (we have to let n go to infinity – there are many technical details to gloss over).

7 Integrability

Many applications lead to situations where you end up with a differential equation on some manifold. Solving these is no easier than it is in the flat case. However, the language of tangent bundles can occasionally make it clearer what is going on, and where the messy formulae actually live.

Furthermore, the existence of solutions to differential equations is essential for showing that the deformations we intuitively are inclined to perform on manifolds actually make sense smoothly. This is reflected in the fact that the flows we construct are smooth.

Example 7.0.1 In the flat case, we are used to drawing "flow charts". For example, given a first-order differential equation

$$\begin{bmatrix} x'(t) \\ y'(t) \end{bmatrix} = f(x(t), y(t))$$

we associate with each point (x, y) the vector $f(x, y)$. In this fashion a first-order ordinary differential equation may be identified with a vector field (Figure 7.1). Each vector would be the velocity vector of a solution to the equation passing through the point (x, y). If f is smooth, the vectors will depend smoothly on the point (it is a smooth vector field), and the picture would resemble a flow of a liquid, where each vector would represent the velocity of the particle at the given point. The paths of each particle would be solutions of the differential equation, and, upon assembling all these solutions, we could talk about the flow of the liquid (Figure 7.2).

7.1 Flows and Velocity Fields

If we are to talk about differential equations on manifolds, the confusion regarding where the velocity fields live (as opposed to the solutions) has to be sorted out. The place of residence of velocity vectors is the tangent bundle, and a differential equation can be represented by a vector field, that is, a section in the tangent bundle $TM \to M$, and its solutions by a "flow".

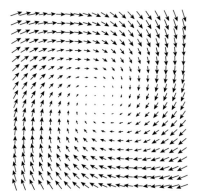

Figure 7.1. The vector field resulting from a system of ordinary differential equations (here a predator–prey system with a stable equilibrium).

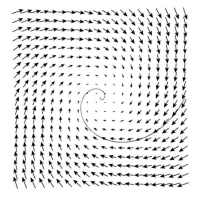

Figure 7.2. A solution to the differential equation is a curve whose derivative equals the corresponding vector field.

Definition 7.1.1 Let M be a smooth manifold. A *(global) flow* on M is a smooth map

$$\Phi \colon \mathbf{R} \times M \to M$$

such that for all $p \in M$ and $s, t \in \mathbf{R}$

- $\Phi(0, p) = p$,
- $\Phi(s, \Phi(t, p)) = \Phi(s + t, p)$.

We are going to show that on a compact manifold there is a one-to-one correspondence between vector fields and global flows. In other words, first-order ordinary differential equations have unique solutions on compact manifolds. This statement is true also for non-compact manifolds, but then we can't expect the flows to be defined on all of $\mathbf{R} \times M$ anymore, and we have to talk about *local flows*. We will return to this later, but first we will familiarize ourselves with global flows.

Our first example is the simplest interesting flow you can think of (one has to admit that the unique flow on $\mathbf{R}^0$ is even simpler, but it isn't particularly interesting). It is so important that we give it in the form of a definition.

Definition 7.1.2 Let $M = \mathbf{R}$, let

$$L: \mathbf{R} \times \mathbf{R} \to \mathbf{R}$$

be the flow given by $L(s, t) = s + t$.

Example 7.1.3 Consider the map

$$\Phi: \mathbf{R} \times \mathbf{R}^2 \to \mathbf{R}^2$$

given by

$$\left(t, \begin{bmatrix} p \\ q \end{bmatrix} \right) \mapsto e^{-t/2} \begin{bmatrix} \cos t & \sin t \\ -\sin t & \cos t \end{bmatrix} \begin{bmatrix} p \\ q \end{bmatrix}.$$

Exercise 7.1.4 Check that this actually **is** a global flow!

For fixed p and q this is the solution to the initial value problem

$$\begin{bmatrix} x' \\ y' \end{bmatrix} = \begin{bmatrix} -1/2 & 1 \\ -1 & -1/2 \end{bmatrix} \begin{bmatrix} x \\ y \end{bmatrix}, \qquad \begin{bmatrix} x(0) \\ y(0) \end{bmatrix} = \begin{bmatrix} p \\ q \end{bmatrix}$$

whose corresponding vector field was used in Figures 7.1 and 7.2 in Example 7.0.1.

A flow is a very structured representation of a vector field.

Definition 7.1.5 Let Φ be a flow on the smooth manifold M. The *velocity field* of Φ is defined to be the vector field

$$\vec{\Phi}: M \to TM,$$

where $\vec{\Phi}(p) = [t \mapsto \Phi(t, p)]$.

The surprise is that *every* (smooth) vector field is the velocity field of a flow (see Theorems 7.2.1 and 7.3.5).

Example 7.1.6 Consider the global flow of Definition 7.1.2. Its velocity field

$$\vec{L}: \mathbf{R} \to T\mathbf{R}$$

is given by $s \mapsto [L_s]$, where L_s is the curve $t \mapsto L_s(t) = L(s, t) = s + t$. Under the standard trivialization $T\mathbf{R} \cong \mathbf{R} \times \mathbf{R}$, $[\omega] \mapsto (\omega(0), \omega'(0))$, we see that $\vec{L}$ is the nonvanishing vector field corresponding to picking out 1 in every fiber.

Example 7.1.7 Consider the flow Φ in Example 7.1.3. Under the standard trivialization $T\mathbf{R}^2 \cong \mathbf{R}^2 \times \mathbf{R}^2$, $[\omega] \mapsto (\omega(0), \omega'(0))$, the velocity field $\overrightarrow{\Phi} \colon \mathbf{R}^2 \to T\mathbf{R}^2$ corresponds to

$$\mathbf{R}^2 \to \mathbf{R}^2 \times \mathbf{R}^2, \qquad \begin{bmatrix} p \\ q \end{bmatrix} \mapsto \left(\begin{bmatrix} p \\ q \end{bmatrix}, \begin{bmatrix} -1/2 & 1 \\ -1 & -1/2 \end{bmatrix} \begin{bmatrix} p \\ q \end{bmatrix} \right).$$

Definition 7.1.8 Let Φ be a global flow on a smooth manifold M, and $p \in M$. The curve

$$\phi_p \colon \mathbf{R} \to M, \qquad \phi_p(t) = \Phi(t, p)$$

is called the *flow line* of Φ through p. The image of the flow line through p is called the *orbit* of p (Figure 7.3).

The orbits split the manifold into disjoint sets.

Exercise 7.1.9 Let $\Phi \colon \mathbf{R} \times M \to M$ be a flow on a smooth manifold M. Then

$$p \sim q \Leftrightarrow \text{ there is a } t \text{ such that } \Phi(t, p) = q$$

defines an equivalence relation on M. Hence, every point in M lies in a unique orbit: different orbits do not intersect.

As an example, consider the flow of Example 7.1.3 (here $M/\sim$ may be identified with $S^1 \coprod \{0\}$).

Example 7.1.10 The flow line through 0 of the flow L of Definition 7.1.2 is the identity on $\mathbf{R}$. The only orbit is $\mathbf{R}$.

More interesting is that the flow lines of the flow of Example 7.1.3 are of two types: the constant flow line at the origin, and the spiraling flow lines filling out the rest of the space.

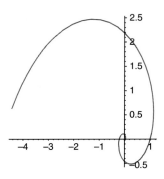

Figure 7.3. The orbit of the point $\begin{bmatrix} 1 \\ 0 \end{bmatrix}$ of the flow of Example 7.1.3.

Exercise 7.1.11   Given $(r, \theta) \in \mathbf{R}^2$ let $\Phi \colon \mathbf{R} \times \mathbf{C} \to \mathbf{C}$ be the flow $\Phi(t, z) = z \cdot r^t e^{it\theta}$. Describe the flow lines when (r, θ) is (i) $(1, 0)$, (ii) $(1, \pi/2)$ and (iii) $(1/2, 0)$.

Note 7.1.12   (This contains important notation, and a reinterpretation of the term "global flow".) Writing $\Phi_t(p) = \Phi(t, p)$ we get another way of expressing a flow. To begin with we have

- $\Phi_0 = $ identity,
- $\Phi_{s+t} = \Phi_s \circ \Phi_t$.

We see that for each t the map Φ_t is a diffeomorphism (with inverse Φ_{-t}) from M to M. The assignment $t \mapsto \Phi_t$ sends sum to composition of diffeomorphisms (i.e., $s + t \mapsto \Phi_{s+t} = \Phi_s \Phi_t$) and defines a continuous "group homomorphism"

$$\mathbf{R} \to \mathrm{Diff}(M),$$

called the *one-parameter family* associated with the flow, from the additive group of real numbers to the group of diffeomorphism (under composition) on M.

We have already used this notation in connection with the flow L of Definition 7.1.2: $L_s(t) = L(s, t) = s + t$.

Definition 7.1.13   Let $\gamma \colon \mathbf{R} \to M$ be a smooth curve on the manifold M. The *velocity vector* $\dot{\gamma}(s) \in T_{\gamma(s)}M$ of γ at $s \in \mathbf{R}$ (Figure 7.4) is defined as the tangent vector

$$\dot{\gamma}(s) = T\gamma \, \overrightarrow{L}(s) = [\gamma L_s] = [t \mapsto \gamma(s + t)].$$

Note 7.1.14   The curve γL_s is given by $t \mapsto \gamma(s + t)$ and $(L_s)'(0) = 1$. So, if (x, U) is a chart with $\gamma(s) \in U$, we get that $\dot{\gamma}(s) \in T_{\gamma(s)}M$ corresponds to $(x\gamma L_s)'(0) = (x\gamma)'(s)$ under the isomorphism $T_{\gamma(s)}M \cong \mathbf{R}^m$ induced by x, explaining the term "velocity vector".

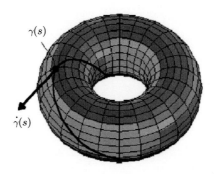

Figure 7.4. The velocity vector $\dot{\gamma}(s)$ of the curve γ at s lives in $T_{\gamma(s)}M$.

The following diagram can serve as a reminder of the definition of the velocity vector and will be used later:

$$
\begin{array}{ccc}
T\mathbf{R} & \xrightarrow{\;T\gamma\;} & TM \\
\scriptstyle{\vec{L}}\big\uparrow & \nearrow{\scriptstyle\dot\gamma} & \big\downarrow \\
\mathbf{R} & \xrightarrow[\;\gamma\;]{} & M.
\end{array}
$$

Definition 7.1.15 Let $X\colon M \to TM$ be a vector field. A *solution curve* is a curve $\gamma\colon J \to M$ (where J is an open interval) such that $\dot\gamma(t) = X(\gamma(t))$ for all $t \in J$.

The velocity field and the flow are intimately connected, and the relation can be expressed in many ways. Here are some. (In this lemma, the equation $\dot\phi_p = \vec\Phi\phi_p$ expressing that "the flow line is a solution to curve to the velocity field" is the one to pay closest attention to.)

Lemma 7.1.16 *Let Φ be a global flow on the smooth manifold M, $p \in M$ and $s \in \mathbf{R}$. Let ϕ_p be the flow line through p and $\Phi_s\colon M \cong M$ the diffeomorphism given by $\phi_p(s) = \Phi_s(p) = \Phi(s, p)$. Then the diagrams*

$$
\begin{array}{ccc}
TM & \xrightarrow[\cong]{\;T\Phi_s\;} & TM \\
\scriptstyle{\vec\Phi}\big\uparrow & & \big\uparrow{\scriptstyle\vec\Phi} \\
M & \xrightarrow[\cong]{\;\Phi_s\;} & M
\end{array}
\qquad
\begin{array}{ccc}
\mathbf{R} & \xrightarrow{\;\phi_p\;} & M \\
\scriptstyle{L_s}\big\downarrow{\scriptstyle\cong} & & {\scriptstyle\cong}\big\downarrow{\scriptstyle\Phi_s} \\
\mathbf{R} & \xrightarrow[\;\phi_p\;]{} & M
\end{array}
\quad and \quad
\begin{array}{ccc}
 & & TM \\
 & \nearrow{\scriptstyle\dot\phi_p} & \big\uparrow{\scriptstyle\vec\Phi} \\
\mathbf{R} & \xrightarrow[\;\phi_p\;]{} & M
\end{array}
$$

commute. Furthermore,

$$
\dot\phi_p(s) = T\Phi_s[\phi_p].
$$

Proof. The claims are variations of the fact that $\Phi(s + t, q) = \Phi(s, \Phi(t, q)) = \Phi(t, \Phi(s, q))$. ∎

Exercise 7.1.17

In this exercise we classify the flow lines of a flow. Let Φ be a flow on a smooth manifold M, and $p \in M$. If $\phi_p\colon \mathbf{R} \to M$ is the flow line of Φ through p, then

- ϕ_p is an injective immersion, or
- ϕ_p is a periodic immersion (i.e., there is a $T > 0$ such that $\phi_p(s) = \phi_p(t)$ if and only if there is an integer k such that $s = t + kT$), or
- ϕ_p is constant.

Note 7.1.18 In the case in which the flow line ϕ_p is a periodic immersion we note that ϕ_p must factor through an imbedding $f\colon S^1 \to M$ with $f(e^{it}) =$

$\phi_p(tT/2\pi)$. That f is an imbedding follows by Corollary 4.7.5, since it is an injective immersion from a compact space.

When ϕ_p is an injective immersion there is no reason to believe that it is an imbedding.

Example 7.1.19   The flow lines in Example 7.1.3 are either constant (the one at the origin) or injective immersions (all the others). The flow

$$\Phi: \mathbf{R} \times \mathbf{R}^2 \to \mathbf{R}^2, \qquad \left(t, \begin{bmatrix} x \\ y \end{bmatrix}\right) \mapsto \begin{bmatrix} \cos t & -\sin t \\ \sin t & \cos t \end{bmatrix} \begin{bmatrix} x \\ y \end{bmatrix}$$

has periodic flow lines (except at the origin).

Exercise 7.1.20   Display an injective immersion $f: \mathbf{R} \to \mathbf{R}^2$ which is not the flow line of a flow.

7.2 Integrability: Compact Case

A difference between vector fields and flows is that vector fields can obviously be added, which makes it easy to custom-build vector fields for a particular purpose. That this is true also for flows is far from obvious, but is one of the nice consequences of the integrability theorem, 7.2.1 below. The theorem allows us to custom-build flows for particular purposes simply by specifying their velocity fields.

Going from flows to vector fields is simple: just take the velocity field. Going the other way is harder, and relies on the fact that first-order ordinary differential equations have unique solutions. We note that the equation

$$\dot{\phi}_p(s) = \vec{\Phi}(\phi_p(s))$$

of Lemma 7.1.16 says that "the flow lines are solution curves to the velocity field". This is the key to proof of the integrability theorem.

Theorem 7.2.1 (Integrability: Compact Case)   *Let M be a smooth compact manifold. Then the velocity field gives a natural bijection between the sets*

$$\{global\ flows\ on\ M\} \leftrightarrows \{vector\ fields\ on\ M\}.$$

Before we prove the integrability theorem, recall the basic existence and uniqueness theorem for ordinary differential equations. For a nice proof giving just continuity see Chapter 5 of Spivak's book [20]. For a complete proof, see, e.g., Chapter IV of Lang's book [12].

Theorem 7.2.2 *Let $f: U \to \mathbf{R}^n$ be a smooth map where $U \subseteq \mathbf{R}^n$ is an open subset and $p \in U$.*

- *(Existence of solution) There is a neighborhood $p \in V \subseteq U$ of p, a neighborhood J of $0 \in \mathbf{R}$ and a smooth map*

$$\Phi: J \times V \to U$$

  such that

  – *$\Phi(0, q) = q$ for all $q \in V$, and*
  – *$\frac{\partial}{\partial t}\Phi(t, q) = f(\Phi(t, q))$ for all $(t, q) \in J \times V$.*

- *(Uniqueness of solution) If γ_1 and γ_2 are smooth curves in U satisfying $\gamma_1(0) = \gamma_2(0) = p$ and*

$$\gamma_i'(t) = f(\gamma_i(t)), \qquad i = 1, 2,$$

  then $\gamma_1 = \gamma_2$ where they both are defined.

Notice that uniqueness gives that Φ satisfies the condition $\Phi(s + t, q) = \Phi(s, \Phi(t, q))$ for small s and t. In more detail, for sufficiently small, but fixed, t let $\gamma_1(s) = \Phi(s + t, q)$ and $\gamma_2(s) = \Phi(s, \Phi(t, q))$. Then $\gamma_1(0) = \gamma_2(0)$ and $\gamma_k'(s) = f(\gamma_k(s))$ for $k = 1, 2$, so $\gamma_1 = \gamma_2$.

Proof. To prove Theorem 7.2.1 we construct an inverse to the function given by the velocity field. That is, given a vector field X on M, we will produce a unique flow Φ whose velocity field is $\vec{\Phi} = X$.

Given a point $p \in M$, choose a chart $x = x_p: U_p \to U_p'$ with $p \in U_p$. Let $H_{x_p}: TU_p \cong U_p' \times \mathbf{R}^n$ be the standard trivialization $H_{x_p}[\gamma] = (x_p\gamma(0), (x_p\gamma)'(0))$. Let $\gamma: J \to U_p$ be a curve and consider the diagram

where $f_p: U_p' \to \mathbf{R}^n$ is defined as the unique map so that the rectangle commutes, i.e., f_p is the composite

$$U_p' \xrightarrow{x_p^{-1}} U_p \xrightarrow{X|_{U_p}} TU_p \xrightarrow{[\gamma] \mapsto (x_p\gamma)'(0)} \mathbf{R}^n,$$

and the commutativity of the triangle says exactly that γ is a solution curve for X on U_p (i.e., that $\dot{\gamma}(t) = X(\gamma(t))$), which we see is equivalent to claiming that

$$(x_p\gamma)'(t) = f_p(x_p\gamma(t)).$$

By the existence and uniqueness theorem for first-order differential equations, Theorem 7.2.2, there is a neighborhood $J_p \times V'_p$ around $(0, x_p(p)) \in \mathbf{R} \times U'_p$ for which there exists a smooth map

$$\Psi = \Psi_p \colon J_p \times V'_p \to U'_p$$

such that $\Psi(0, q) = q$ for all $q \in V'_p$ and $\frac{\partial}{\partial t}\Psi(t, q) = f_p(\Psi(t, q))$ for all $(t, q) \in J_p \times V'_p$. Furthermore, for each $q \in V'_p$ the curve $\Psi(-, q) \colon J_p \to U'_p$ is unique with respect to this property.

The set of open sets of the form $V_p = x_p^{-1} V'_p$ is an open cover of M, and hence we may choose a finite subcover. Let J be the intersection of the J_ps corresponding to this finite cover. Since it is a finite intersection, J contains an open interval $(-\epsilon, \epsilon)$ around 0.

What happens to f_p when we vary p? Let p' be another point, $q \in U = V_p \cap V_{p'}$, and consider the commutative diagram (restrictions suppressed)

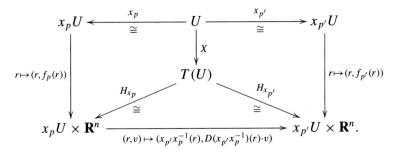

Hence, we get that $f_{p'}x_{p'}x_p^{-1}(r) = D(x_{p'}x_p^{-1})(r) \cdot f_p(r)$ for $r \in x_p U$. So, if we set $P(t, q) = x_{p'}x_p^{-1}\Psi_p(t, x_p x_{p'}^{-1}(q))$, the flat chain rule Lemma 3.0.3 gives that $\frac{\partial}{\partial t}P(t, q) = f_{p'}(P(t, q))$. Since in addition $P(0, q) = q$, we get that both P and $\Psi_{p'}$ are solutions to the initial value problem (with $f_{p'}$), and by uniqueness $P = \Psi_{p'}$ on the domain of definition. In other words,

$$x_p^{-1}\Psi_p(t, x_p(q)) = x_{p'}^{-1}\Psi_{p'}(t, x_{p'}(q)), \qquad q \in U, \quad t \in J.$$

Hence we may define a smooth map

$$\tilde{\Phi} \colon J \times M \to M$$

by $\tilde{\Phi}(t, q) = x_p^{-1}\Psi_p(t, x_p q)$ if $q \in x_p^{-1}V'_p$. Note that the uniqueness of the solution also gives that $\tilde{\Phi}(t, \tilde{\Phi}(s, q)) = \tilde{\Phi}(s + t, q)$ for $|s|, |t|$ and $|s + t|$ less than ϵ. This also means that we may extend the domain of definition to get a map

$$\Phi \colon \mathbf{R} \times M \to M,$$

since for any $t \in \mathbf{R}$ there is a natural number k such that $|t/k| < \epsilon$, and we simply define $\Phi(t, q)$ as $\tilde{\Phi}_{t/k}$ applied k times to q. ∎

The condition that M was compact was crucial to this proof. A similar statement is true for non-compact manifolds, and we will return to that statement later.

Exercise 7.2.3 You are given two flows Φ_N and Φ_S on the sphere S^2. Why does there exist a flow Φ with $\Phi(t, q) = \Phi_N(t, q)$ for small t and q close to the North pole, and $\Phi(t, q) = \Phi_S(t, q)$ for small t and q close to the South pole?

Exercise 7.2.4 Construct a vector field on the torus such that the solution curves are imbedded circles. Construct a vector field on the torus such that the solution curves are dense immersions.

Exercise 7.2.5 Let $O(n)$ be the orthogonal group, and recall from Exercise 5.5.20 the isomorphism between the tangent bundle of $O(n)$ and the projection on the first factor

$$E = \{(g, A) \in O(n) \times M_n(\mathbf{R}) \mid A^T = -g^T A g^T\} \to O(n).$$

Choose a skew matrix $A \in M_n(\mathbf{R})$ (i.e., such that $A^T = -A$), and consider the vector field $X \colon O(n) \to TO(n)$ induced by

$$O(n) \to E,$$
$$g \mapsto (g, gA).$$

Show that the flow associated with X is given by $\Phi(s, g) = g e^{sA}$, where the exponential is defined as usual by $e^B = \sum_{n=0}^{\infty} B^n / n!$.

7.3 Local Flows and Integrability

We now make the modifications necessary for proving an integrability theorem also in the non-compact case. On manifolds that are not compact, the concept of a (global) flow is not the correct one. This can be seen by considering a global flow Φ on some manifold M and restricting it to some open submanifold U. Then some of the flow lines may leave U after finite time. To get a "flow" Φ_U on U we must then accept that Φ_U is defined only on some open subset of $\mathbf{R} \times U$ containing $\{0\} \times U$.

Also, if we jump ahead a bit, and believe that flows should correspond to general solutions to first-order ordinary differential equations (that is, vector fields), you may consider the initial value problem $y' = y^2$, $y(0) = y_0$ on $M = \mathbf{R}$ (the corresponding vector field is $\mathbf{R} \to T\mathbf{R}$ given by $s \mapsto [t \mapsto s + s^2 t]$).

The solution to $y' = y^2$, $y(0) = y_0$ is

$$y(t) = \begin{cases} \dfrac{1}{1/y_0 - t} & \text{if } y_0 \neq 0 \\ 0 & \text{if } y_0 = 0 \end{cases}$$

and the domain (Figure 7.5) of the "flow"

$$\Phi(t, p) = \begin{cases} \dfrac{1}{1/p - t} & \text{if } p \neq 0 \\ 0 & \text{if } p = 0 \end{cases}$$

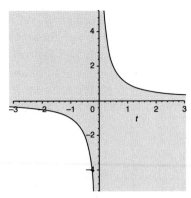

Figure 7.5. The domain A of the "flow". It contains an open neighborhood around $\{0\} \times M$.

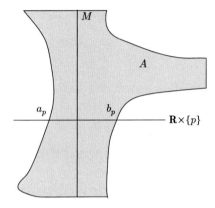

Figure 7.6. The "horizontal slice" J_p is an open interval (a_p, b_p) containing zero.

is

$$A = \{(t, p) \in \mathbf{R} \times \mathbf{R} \mid pt < 1\}.$$

Definition 7.3.1 Let M be a smooth manifold. A *local flow* is a smooth map $\Phi \colon A \to M$, where

$$A \subseteq \mathbf{R} \times M$$

is open and contains $\{0\} \times M$, such that for each $p \in M$

$$J_p \times \{p\} = A \cap (\mathbf{R} \times \{p\})$$

is connected (so that J_p is an open interval (a_p, b_p) containing 0; Figure 7.6) and such that for all s, t, p with (t, p), $(s + t, p)$ and $(s, \Phi(t, p))$ in A

- $\Phi(0, p) = p$,
- $\Phi(s, \Phi(t, p)) = \Phi(s + t, p)$.

Definition 7.3.2   A local flow $\Phi\colon A \to M$ is *maximal* if there is no local flow $\Psi\colon B \to M$ such that $A \subsetneq B$ and $\Psi|_A = \Phi$.

Note 7.3.3   The definitions of the *velocity field*

$$\overrightarrow{\Phi}\colon M \to TM$$

(the tangent vector $\overrightarrow{\Phi}(p) = [t \mapsto \Phi(t, p)]$ depends only on the values of the curve in a small neighborhood of 0), the *flow line*

$$\phi_p\colon J_p \to M, \qquad t \mapsto \phi_p(t) = \Phi(t, p)$$

and the *orbit*

$$\phi_p(J_p) \subseteq M$$

through $p \in M$ make just as much sense for a local flow Φ.

However, we can't talk about "the diffeomorphism Φ_t," since there may be $p \in M$ such that $(t, p) \notin A$, and so Φ_t is not defined on all of M.

Example 7.3.4   Check that the formula

$$\Phi(t, p) = \begin{cases} \dfrac{1}{1/p - t} & \text{if } p \neq 0 \\ 0 & \text{if } p = 0 \end{cases}$$

defines a local flow with velocity field $\overrightarrow{\Phi}\colon \mathbf{R} \to T\mathbf{R}$ given by $s \mapsto [t \mapsto \Phi(t, s)]$ (which under the standard trivialization

$$T\mathbf{R} \xrightarrow{\;[\omega] \mapsto (\omega(0), \omega'(0))\;} \mathbf{R} \times \mathbf{R}$$

corresponds to $s \mapsto (s, s^2)$ – and so $\overrightarrow{\Phi}(s) = [t \mapsto \Phi(t, s)] = [t \mapsto s + s^2 t]$) with domain

$$A = \{(t, p) \in \mathbf{R} \times \mathbf{R} \mid pt < 1\}$$

and so $a_p = 1/p$ for $p < 0$ and $a_p = -\infty$ for $p \geq 0$. Note that Φ_t only makes sense for $t = 0$.

Theorem 7.3.5 (Integrability)   *Let M be a smooth manifold. Then the velocity field gives a natural bijection between the sets*

$$\{\textit{maximal local flows on } M\} \leftrightarrows \{\textit{vector fields on } M\}.$$

Proof. The essential idea is the same as in the compact case, but we have to worry a bit more about the domain of definition of our flow. The local solution to the ordinary differential equation means that we have unique maximal solution curves

$$\phi_p \colon J_p \to M$$

for all p. This also means that the curves $t \mapsto \phi_p(s+t)$ and $t \mapsto \phi_{\phi_p(s)}(t)$ agree (both are solution curves through $\phi_p(s)$), and we define

$$\Phi \colon A \to M$$

by setting

$$A = \bigcup_{p \in M} J_p \times \{p\}, \text{ and } \Phi(t, p) = \phi_p(t).$$

The only questions are whether A is open and Φ is smooth. But this again follows from the local existence theorems: around any point in A there is a neighborhood on which Φ corresponds to the unique local solution (see pages 82 and 83 of [4] for more details). ■

Note that maximal flows that are not global must leave any given compact subset within finite time.

Lemma 7.3.6 *Let $K \subset M$ be a compact subset of a smooth manifold M, and let Φ be a maximal local flow on M such that $b_p < \infty$. Then there is an $\epsilon > 0$ such that $\Phi(t, p) \notin K$ for $t > b_p - \epsilon$.*

In particular, the maximal local flow associated with a vector field vanishing outside a compact set is global.

Proof. Since K is compact there is an $\epsilon > 0$ such that

$$[-\epsilon, \epsilon] \times K \subseteq A \cap (\mathbf{R} \times K)$$

(cover $\{0\} \times K$ by open sets of the form $(-\delta, \delta) \times U$ in the open set A, choose a finite subcover and let ϵ be less than the minimum δ). The problem is then that "you can flow ϵ more from *anywhere* in K": if there is a $T \in (b_p - \epsilon, b_p)$ such that $\Phi(T, p) \in K$ we could extend Φ by setting $\Phi(t, p) = \Phi(t - T, \Phi(T, p))$ for all $t \in [T, T + \epsilon]$, contradicting the maximality of b_p (since $b_p < T + \epsilon$). ■

Here is a particularly important flow afforded by the integrability theorem. It will reappear at a crucial point in our discussion of Morse theory and is generally used whenever you want to deform a submanifold in a controlled fashion.

Definition 7.3.7 Let M be a Riemannian manifold and let $f \colon M \to \mathbf{R}$ be a smooth function. The *gradient flow* is the local flow on M associated through the integrability theorem, 7.3.5, with the gradient field grad $f : M \rightarrow TM$ of Definition 6.6.8.

Example 7.3.8 The function $f : \mathbf{R} \rightarrow \mathbf{R}$ given by $f(s) = s^2$ of Example 5.6.3 has gradient $\text{grad}_s f = (s, [t \mapsto s + s^2 t])$ (the Jacobi matrix is $Df(s) = 2s$) and so the gradient flow is the local flow of Exercise 7.3.4.

Note 7.3.9 Some readers may worry about the fact that we do not consider "time-dependent" differential equations, but, by a simple trick as on page 226 in [20], these are covered by our present considerations.

Exercise 7.3.10 Find a nonvanishing vector field on $\mathbf{R}$ whose solution curves are defined only on finite intervals.

7.4 Second-Order Differential Equations[1]

We give a brief and inadequate sketch of second-order differential equations. This is important for a wide variety of applications, in particular for the theory of geodesics which will be briefly discussed in Section 8.2.7 after partitions of unity have been introduced.

For a smooth manifold M let $\pi_M : TM \rightarrow M$ be the tangent bundle (we just need a decoration on π to show its dependence on M).

Definition 7.4.1 A *second-order differential equation* on a smooth manifold M is a smooth map

$$\xi : TM \rightarrow TTM$$

such that

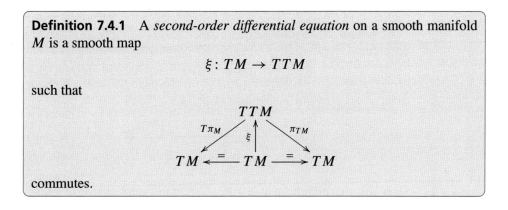

commutes.

Note 7.4.2 The equality $\pi_{TM}\xi = \text{id}_{TM}$ just says that ξ is a vector field on TM; it is the other equality, $(T\pi_M)\xi = \text{id}_{TM}$, which is crucial.

---

[1] This section is not referred to later in the book except in the example on the exponential map, Example 8.2.7.

Exercise 7.4.3

The flat case: the reference sheet (c.f. Exercise 5.5.25). Make sense of the following remarks and write down your interpretation.

A curve in TM is an equivalence class of "surfaces" in M, for if $\beta\colon J \to TM$ then for each $t \in J$ we have that $\beta(t)$ must be an equivalence class of curves, $\beta(t) = [\omega(t)]$, and we may think of $t \mapsto \{s \mapsto \omega(t)(s)\}$ as a surface if we let s and t move simultaneously. In more detail, on choosing a chart we may assume that our manifold is an open set U in $\mathbf{R}^n$, and then we have the trivialization

$$TU \xrightarrow[\cong]{[\omega] \mapsto (\omega(0),\omega'(0))} U \times \mathbf{R}^n$$

with inverse $(p, v) \mapsto [t \mapsto p + tv]$ (the directional derivative at p in the vth direction) and

$$T(TU) \xrightarrow[\cong]{[\beta] \mapsto (\beta(0),\beta'(0))} T(U) \times (\mathbf{R}^n \times \mathbf{R}^n)$$

$$\xrightarrow[\cong]{(\beta(0),\beta'(0)) \mapsto ((\omega(0,0),D_2\omega(0,0)),(D_1\omega(0,0),D_2D_1\omega(0,0)))} (U \times \mathbf{R}^n) \times (\mathbf{R}^n \times \mathbf{R}^n)$$

with inverse $(p, v_1, v_2, v_3) \mapsto [t \mapsto [s \mapsto \omega(t)(s)]]$ with

$$\omega(t)(s) = p + sv_1 + tv_2 + stv_3.$$

Hence, if $\gamma\colon J \to U$ is a curve, then $\dot\gamma$ corresponds to the curve

$$J \xrightarrow{t \mapsto (\gamma(t),\gamma'(t))} U \times \mathbf{R}^n;$$

and if $\beta\colon J \to TU$ corresponds to $t \mapsto (x(t), v(t))$, then $\dot\beta$ corresponds to

$$J \xrightarrow{t \mapsto (x(t),v(t),x'(t),v'(t))} U \times \mathbf{R}^n \times \mathbf{R}^n \times \mathbf{R}^n.$$

Hence, the two equations $\pi_{TM}\xi = \mathrm{id}_{TM}$ and $(T\pi_M)\xi = \mathrm{id}_{TM}$ for a second-order differential equation $\xi\colon TU \to TTU$ give that ξ corresponds to a map

$$U \times \mathbf{R}^n \xrightarrow{(p,v) \mapsto (p,v,v,f(p,v))} U \times \mathbf{R}^n \times \mathbf{R}^n \times \mathbf{R}^n.$$

This means that $\ddot\gamma = \dot{\dot\gamma}$ corresponds to

$$J \xrightarrow{t \mapsto (\gamma(t),\gamma'(t),\gamma'(t),\gamma''(t))} U \times \mathbf{R}^n \times \mathbf{R}^n \times \mathbf{R}^n.$$

Exercise 7.4.4

Show that our definition of a second-order differential equation corresponds to the usual notion of a second-order differential equation in the case $M = \mathbf{R}^n$.

Definition 7.4.5 Given a second-order differential equation

$$\xi\colon TM \to TTM,$$

a curve $\gamma\colon J \to M$ is called a *solution curve* for ξ on M if $\dot\gamma$ is a solution curve to ξ "on TM".

Note 7.4.6   Spelling this out, we have that

$$\ddot{\gamma}(t) = \xi(\dot{\gamma}(t))$$

for all $t \in J$. Note the bijection

{solution curves $\beta \colon J \to TM$} $\leftrightarrows$ {solution curves $\gamma \colon J \to M$}, $\begin{array}{l} \dot{\gamma} \leftarrow \gamma, \\ \beta \mapsto \pi_M \beta. \end{array}$

Exercise 7.4.7   The inclusion $S^n \subseteq \mathbf{R}^{n+1}$ induces isomorphisms $TS^n \cong E$ and $TTS^n \cong F$, where $E = \{(p, v) \in S^n \times \mathbf{R}^{n+1} \mid p \cdot v = 0\}$ and

$$F = \Big\{ (p, v_1, v_2, v_3) \in S^n \times \mathbf{R}^{n+1} \times \mathbf{R}^{n+1} \times \mathbf{R}^{n+1} \mid \begin{smallmatrix} p \cdot v_1 = p \cdot v_2 = \\ v_1 \cdot v_2 + p \cdot v_3 = 0 \end{smallmatrix} \Big\}.$$

Show that the great circle given by

$$\gamma(t) = \cos(|v|t)p + \frac{\sin(|v|t)}{|v|} v$$

defines a solution curve for the second-order differential equation $\xi \colon TS^n \to TTS^n$ induced by

$$E \xrightarrow{\ (p,v) \mapsto (p,v,v,-|v|^2 p)\ } F.$$

Exercise 7.4.8   Show that the set of second-order differential equations on a smooth manifold M is a *convex subset* of the vector space of all vector fields $\xi \colon TM \to TTM$; that is, if $s_1 + s_2 = 1$ and ξ_1 and ξ_2 are second-order differential equations, then $s_1 \xi_1 + s_2 \xi_2$ is a second-order differential equation.

8 Local Phenomena that Go Global

In this chapter we define partitions of unity. They are smooth devices making it possible to patch together some types of local information into global information. They come in the form of "bump functions" such that around any given point there are only finitely many of them that are nonzero, and such that the sum of their values is 1.

This can be applied for instance to patch together the nice local structure of a manifold to an imbedding into a Euclidean space (we do it in the compact case, see Theorem 8.2.6), to construct sensible metrics on the tangent spaces (so-called Riemannian metrics, see Section 6.6), and in general to construct smooth functions with desirable properties. We will also use it to prove Ehresmann's fibration theorem, 8.5.10.

8.1 Refinements of Covers

In order to patch local phenomena together, we will be using the fact that manifolds can be covered by chart domains in a very orderly fashion, by means of what we will call "good" atlases. This section gives the technical details needed.

If $0 < r$ let $E^n(r) = \{x \in \mathbf{R}^n \mid |x| < r\}$ be the open n-dimensional ball of radius r centered at the origin.

Lemma 8.1.1 *Let M be an n-dimensional manifold. Then there is a countable atlas $\mathcal{A}$ such that $x(U) = E^n(3)$ for all $(x, U) \in \mathcal{A}$ and such that*

$$\bigcup_{(x,U)\in\mathcal{A}} x^{-1}(E^n(1)) = M.$$

If M is smooth, then all charts may be chosen to be smooth.

Proof. Let $\mathcal{B}$ be a countable basis for the topology on M. For every $p \in M$ there is a chart (x, U) with $x(p) = 0$ and $x(U) = E^n(3)$. The fact that $\mathcal{B}$ is a basis for the topology gives that there is a $V \in \mathcal{B}$ with

$$p \in V \subseteq x^{-1}(E^n(1)).$$

For each such $V \in \mathcal{B}$ choose just **one** such chart (x, U) with $x(U) = E^n(3)$ and

$$x^{-1}(0) \in V \subseteq x^{-1}(E^n(1)).$$

The set of these charts is the desired countable $\mathcal{A}$. If M were smooth, we would just insert "smooth" in front of every "chart" in the proof above.   ∎

Lemma 8.1.2   *Let M be a manifold. Then there is a sequence $A_1 \subseteq A_2 \subseteq \ldots$ of compact subsets of M such that for every $i \geq 1$ the compact subset A_i is contained in the interior of A_{i+1} and such that $\bigcup_i A_i = M$.*

Proof.   Let $\{(x_i, U_i)\}_{i=1,\ldots}$ be the countable atlas of the lemma above, and let

$$A_k = \bigcup_{i=1}^{k} x_i^{-1}(\overline{E^n(2 - 1/k)}).$$ ∎

Definition 8.1.3   Let $\mathcal{U}$ be an open cover of a space X. We say that another cover $\mathcal{V}$ is a *refinement* of $\mathcal{U}$ if every member of $\mathcal{V}$ is contained in a member of $\mathcal{U}$.

Definition 8.1.4   Let $\mathcal{U}$ be an open cover of a space X. We say that $\mathcal{U}$ is *locally finite* if each $p \in X$ has a neighborhood which intersects only finitely many sets in $\mathcal{U}$.

Definition 8.1.5   Let M be a manifold and let $\mathcal{U}$ be an open cover of M. A *good atlas* subordinate to $\mathcal{U}$ is a countable atlas $\mathcal{A}$ on M such that

(1) the cover $\{V\}_{(x,V)\in\mathcal{A}}$ is a locally finite refinement of $\mathcal{U}$,
(2) $x(V) = E^n(3)$ for each $(x, V) \in \mathcal{A}$, and
(3) $\bigcup_{(x,V)\in\mathcal{A}} x^{-1}(E^n(1)) = M$.

Theorem 8.1.6   *Let M be a manifold and let $\mathcal{U}$ be an open cover of M. Then there exists a good atlas $\mathcal{A}$ subordinate to $\mathcal{U}$. If M is smooth, then $\mathcal{A}$ may be chosen to be a subatlas of the smooth structure.*

Proof.   By Lemma 8.1.2 we may choose a sequence

$$A_1 \subseteq A_2 \subseteq A_3 \subseteq \ldots$$

of compact subsets of M such that for every $i \geq 1$ the compact subset A_i is contained in the interior of A_{i+1} and such that $\bigcup_i A_i = M$. For every point

$$p \in A_{i+1} - \text{int}(A_i)$$

choose a $U_p \in \mathcal{U}$ with $p \in U_p$ and a (smooth) chart (y_p, W_p) such that $p \in W_p \subseteq U_p$ and $y_p(p) = 0$. Since W_P and $\text{int}(A_{i+2}) - A_{i-1}$ are open there is an $\epsilon_p > 0$ such that

$$E^n(\epsilon_p) \subseteq y_p((\text{int}(A_{i+2}) - A_{i-1}) \cap W_p).$$

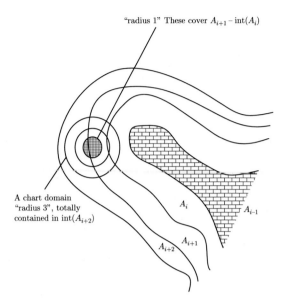

Figure 8.1. The positioning of the charts.

Let $V_p = y_p^{-1}(E^n(\epsilon_p))$ and

$$x_p = \frac{3}{\epsilon_p} y_p|_{V_p} : V_p \to E^n(3).$$

Then $\{x_p^{-1}(E^n(1))\}_p$ covers the compact set $A_{i+1} - \mathrm{int}(A_i)$, and we may choose a finite set of points $p_1, \ldots, p_k$ such that

$$\{x_{p_j}^{-1}(E^n(1))\}_{j=1,\ldots,k}$$

still covers $A_{i+1} - \mathrm{int}(A_i)$. See Figure 8.1.

Letting $\mathcal{A}$ consist of the (x_{p_j}, V_{p_j}) as i and j vary we have proven the theorem.

■

8.2 Partition of Unity

Definition 8.2.1   A family of continuous functions

$$\{\phi_\alpha : X \to [0, 1] \mid \alpha \in A\}$$

is called a *partition of unity* if

$\{\phi_\alpha^{-1}((0, 1]) \mid \alpha \in A\}$ is a locally finite (Definition 8.1.4) open cover of X and for each $p \in X$ the (finite) sum $\sum_\alpha \phi_\alpha(p) = 1$.

The partition of unity is said to be *subordinate* to a cover $\mathcal{U}$ of X if in addition for each ϕ_α there is a $U \in \mathcal{U}$ with $\mathrm{supp}(\phi_\alpha) = \overline{\phi_\alpha^{-1}((0, 1])} \subseteq U$.

Given a space that is not too big and complicated (for instance, if it is a compact manifold), it may not be surprising that we can build a partition of unity on it. What is more surprising is that on smooth manifolds we can build **smooth** partitions of unity (that is, considered as real-valued functions, all the ϕ_αs are smooth).

In order to do this we need smooth bump functions. In particular, we will use the smooth bump function

$$\gamma_{(1,1)} \colon \mathbf{R}^n \to \mathbf{R}$$

defined in Lemma 3.2.3, which has the properties that

- $\gamma_{(1,1)}(p) = 1$ for all p with $|p| \le 1$,
- $\gamma_{(1,1)}(p) \in (0,1)$ for all p with $|p| \in (1,2)$, and
- $\gamma_{(1,1)}(p) = 0$ for all $p \in \mathbf{R}^n$ with $|p| \ge 2$.

Theorem 8.2.2 *Let M be a smooth manifold, and let $\mathcal{U}$ be a cover of M. Then there is a smooth partition of unity of M subordinate to $\mathcal{U}$.*

Proof. To the good atlas $\mathcal{A} = \{(x_i, V_i)\}$ subordinate to $\mathcal{U}$ constructed in Theorem 8.1.6 we may assign functions $\{\psi_i\}$ as follows:

$$\psi_i(q) = \begin{cases} \gamma_{(1,1)}(x_i(q)) & \text{for } q \in V_i = x_i^{-1}(E^n(3)) \\ 0 & \text{otherwise.} \end{cases}$$

The function ψ_i has support $\overline{x_i^{-1}(E^n(2))}$ and is obviously smooth. Since $\{V_i\}$ is locally finite, around any point $p \in M$ there is an open set such that there are only finitely many ψ_is with nonzero values, and hence the expression

$$\sigma(p) = \sum_i \psi_i(p)$$

defines a smooth function $M \to \mathbf{R}$ with everywhere positive values. The partition of unity is then defined by $\phi_i(p) = \psi_i(p)/\sigma(p)$. ∎

Exercise 8.2.3

Let M be a smooth manifold, $f \colon M \to \mathbf{R}$ a continuous function and ϵ a positive real number. Then there is a smooth $g \colon M \to \mathbf{R}$ such that for all $p \in M$

$$|f(p) - g(p)| < \epsilon.$$

You may use without proof Weierstrass' theorem, which says the following. Suppose that $f \colon K \to \mathbf{R}$ is a continuous function with $K \subseteq \mathbf{R}^m$ compact. For every $\epsilon > 0$, there exists a polynomial g such that for all $x \in K$ we have $|f(x) - g(x)| < \epsilon$.

Exercise 8.2.4

Let $L \to M$ be a smooth line bundle. Show that $L \otimes L \to M$ is trivial.

Example 8.2.5 (Imbeddings in Euclidean Space)   As an application of bump functions, we will prove an easy version of Whitney's imbedding theorem. The hard version states that any manifold may be imbedded in the Euclidean space of the double dimension. As a matter of fact, with the appropriate topology, the space of imbeddings $M \to \mathbf{R}^{2n+1}$ is dense in the space of all smooth maps $M \to \mathbf{R}^{2n+1}$ (see, e.g., Section 2.1.0 of [8], or the more refined version, Section 2.2.13 of [8]). We will prove only the following theorem.

Theorem 8.2.6   *Let M be a compact smooth manifold. Then there is an imbedding $M \to \mathbf{R}^N$ for some N.*

Proof.   (After [4]) Assume M has dimension m. Choose a finite good atlas

$$\mathcal{A} = \{x_i, V_i\}_{i=1,\dots,r}.$$

Define $\psi_i : M \to \mathbf{R}$ and $k_i : M \to \mathbf{R}^m$ by

$$\psi_i(p) = \begin{cases} \gamma_{(1,1)}(x_i(p)) & \text{for } p \in V_i \\ 0 & \text{otherwise,} \end{cases}$$

$$k_i(p) = \begin{cases} \psi_i(p) \cdot x_i(p) & \text{for } p \in V_i \\ 0 & \text{otherwise.} \end{cases}$$

Consider the map

$$f : M \to \prod_{i=1}^{r} \mathbf{R}^m \times \prod_{i=1}^{r} \mathbf{R},$$
$$p \mapsto ((k_1(p), \dots, k_r(p)), (\psi_1(p), \dots, \psi_r(p))).$$

Using that M is compact, we shall prove that f is an imbedding by showing that it is an injective immersion (c.f. Corollary 4.7.5).

First, f is an immersion, because for every $p \in M$ there is a j such that $T_p k_j$ has rank m.

Secondly, assume $f(p) = f(q)$ for two points $p, q \in M$. Assume $p \in x_j^{-1}(E^m(1))$. Then we must have that q is also in $x_j^{-1}(\overline{E^m(1)})$ (since $\psi_j(p) = \psi_j(q)$). But then we have that $k_j(p) = x_j(p)$ is equal to $k_j(q) = x_j(q)$, and hence $p = q$ since x_j is a bijection. ∎

Techniques like this are used to construct imbeddings. However, occasionally it is important to know when imbeddings are not possible, and then these techniques are of no use. For instance, why can't we imbed $\mathbf{RP}^2$ in $\mathbf{R}^3$? Proving this directly is probably quite hard, and for such problems algebraic topology is useful.

Example 8.2.7 (the Exponential Map)[1]   As an example of the use of partitions of unity, we end this section with a string of exercises leading to a quick definition of the exponential map from the tangent space to the manifold.

Exercise 8.2.8

(The Existence of "Geodesics") The second-order differential equation $T\mathbf{R}^n \to TT\mathbf{R}^n$ corresponding to the map

$$\mathbf{R}^n \times \mathbf{R}^n \to \mathbf{R}^n \times \mathbf{R}^n \times \mathbf{R}^n \times \mathbf{R}^n, \qquad (x, v) \mapsto (x, v, v, 0)$$

has solution curves given by the straight line $t \mapsto x + tv$ (a straight curve has zero second derivative). Prove that you may glue together these straight lines by means of charts and partitions of unity to get a second-order differential equation (see Definition 7.4.1)

$$\xi : TM \to TTM$$

with the *spray* property that for all $s \in \mathbf{R}$

$$
\begin{array}{ccc}
TTM & \xrightarrow{\ Ts\ } & TTM \\[4pt]
{\scriptstyle s\xi}\uparrow & & \uparrow{\scriptstyle \xi} \\[4pt]
TM & \xrightarrow{\ s\ } & TM
\end{array}
$$

commutes, where $s : TM \to TM$ is multiplication by s in each fiber. A solution γ of the second-order differential equation ξ is then called a geodesic with respect to ξ with initial condition $\dot{\gamma}(0)$.

Note 8.2.9   The significance of the spray property in Exercise 8.2.8 is that "you may speed up (by a factor s) along a geodesic, but the orbit won't change" (see Exercise 8.2.10). Second-order differential equations satisfying the spray property are simply referred to as *sprays*.

For instance, under the trivializations given in Exercise 8.2.8, that a vector field on $T\mathbf{R}^n$ has the spray property corresponds to saying that the map

$$\mathbf{R}^n \times \mathbf{R}^n \to \mathbf{R}^n \times \mathbf{R}^n \times \mathbf{R}^n \times \mathbf{R}^n, \qquad (p, v) \mapsto (p, v, f_1(p, v), f_2(p, v))$$

has the property that, if $s \in \mathbf{R}$, then $f_1(p, sv) = sf_1(p, v)$ and $f_2(p, sv) = s^2 f_2(p, v)$. The condition that it is a second-order differential equation requires in addition that $f_1(p, v) = v$.

Exercise 8.2.10

Let ξ be a spray on a smooth manifold M, let $\gamma : (a, b) \to M$ be a geodesic with respect to ξ and let s be a nonzero real number. Then the curve $\gamma_s = \gamma s : (a/s, b/s) \to M$ (interpreted properly if s is negative) given by $\gamma_s(t) = \gamma(st)$ is a geodesic with initial condition $s\dot{\gamma}(0)$.

----

[1] Geodesics and the exponential map are important for many applications, but are not used later on, so the rest of this section may be skipped without disrupting the flow.

Exercise 8.2.11

(Definition of the Exponential Map) Given a second-order differential equation $\xi \colon TM \to TTM$ as in Exercise 8.2.8, consider the corresponding local flow $\Phi \colon A \to TM$, define the open neighborhood of the zero section

$$\mathcal{T} = \{[\omega] \in TM \mid 1 \in A \cap (\mathbf{R} \times \{[\omega]\})\}$$

and you may define the *exponential map*

$$\exp \colon \mathcal{T} \to M$$

by sending $[\omega] \in TM$ to $\pi_M \Phi(1, [\omega])$.

Essentially exp says the following: for a tangent vector $[\omega] \in TM$ start out in $\omega(0) \in M$ in the direction on $\omega'(0)$ and travel a unit in time along the corresponding geodesic. This "wraps the tangent space of a point p down to the manifold", making it possible to identify small tangent vectors with points near p (just as $[\gamma] \mapsto p + \gamma'(0)$ gives an identification $T_p\mathbf{R}^n \cong \mathbf{R}^n$).

The exponential map depends on ξ. Alternatively we could have given a definition of exp using a choice of a Riemannian metric, which would be more in line with the usual treatment in differential geometry.

Exercise 8.2.12

Show that the second-order differential equation $\xi \colon TS^n \to TTS^n$ defined in Exercise 7.4.7 is a spray, i.e., it satisfies the equation $Ts(s\xi) = \xi s$ as in Exercise 8.2.8. Since we showed in Exercise 7.4.7 that the great circles $\gamma(t) = \cos(|v|t)\,p + (\sin(|v|t)/|v|)v$ are solution curves we get that the great circles are geodesics with respect to ξ.

Show that, with respect to this spray, the exponential $\exp \colon TS^n \to S^n$ is given by

$$\exp([v]) = \cos(|v'(0)|)v(0) + \frac{\sin(|v'(0)|)}{|v'(0)|}\,v'(0).$$

8.3 Global Properties of Smooth Vector Bundles

In Section 6.6 we studied how the presence of a fiber metric on a vector bundle simplified many problems. Using partitons of unity, we now show that *all* smooth vector bundes support fiber metrics, so that these simplifications actually hold in total generality.

Theorem 8.3.1 *Let M be a smooth manifold and let $E \to M$ be a smooth bundle. Then there is a smooth fiber metric on $E \to M$. In particular, any smooth manifold supports a Riemannian metric.*

Proof. Assume the rank of E is n. Let $\mathcal{B}$ be the bundle atlas. Choose a good atlas $\mathcal{A} = \{(x_i, V_i)\}_{i \in \mathbf{N}}$ subordinate to $\{U \mid (h, U) \in \mathcal{B}\}$ and a smooth partition of unity $\{\phi_i \colon M \to \mathbf{R}\}$ with $\mathrm{supp}(\phi_i) \subset V_i$ as given by the proof of Theorem 8.2.2.

Since for any of the V_is, there is a bundle chart (h, U) in $\mathcal{B}$ such that $V_i \subseteq U$, the bundle restricted to any V_i is trivial. Hence we may choose a fiber metric, i.e., a section

$$\sigma_i : V_i \to SB(E)|_{V_i}$$

such that σ_i is (bilinear, symmetric and) positive definite on every fiber. For instance we may let $\sigma_i(p) \in SB(E_p)$ be the positive definite symmetric bilinear map

$$E_p \times E_p \xrightarrow{h_p \times h_p} \mathbf{R}^n \times \mathbf{R}^n \xrightarrow{(v,w) \mapsto v \cdot w = v^{\mathrm{T}} w} \mathbf{R}.$$

Let $g_i : M \to SB(E)$ be defined by

$$g_i(p) = \begin{cases} \phi_i(p)\sigma_i(p) & \text{if } p \in V_i \\ 0 & \text{otherwise}, \end{cases}$$

and let $g : M \to SB(E)$ be given as the sum $g(p) = \sum_i g_i(p)$. The property "positive definite" is *convex*, i.e., if σ_1 and σ_2 are two positive definite forms on a vector space and $t \in [0, 1]$, then $t\sigma_1 + (1 - t)\sigma_2$ is also positive definite (since $t\sigma_1(v, v) + (1 - t)\sigma_2(v, v)$ must obviously be non-negative, and can be zero only if $\sigma_1(v, v) = \sigma_2(v, v) = 0$). By induction we get that $g(p)$ is positive definite since all the $\sigma_i(p)$s were positive definite (Figure 8.2). ∎

On combining Theorem 8.3.1 with point 4 of Lemma 6.6.13 we get the following convenient corollary.

Corollary 8.3.2 *Every smooth vector bundle possesses an atlas whose transition functions map to the orthogonal group.*

Note 8.3.3 This is an example of the notion of *reduction of the structure group*, in this case from $\mathrm{GL}_n(\mathbf{R})$ to $O(n)$; the corollary says that this is always possible.

Usually a reduction of the structure group tells us something important about the bundle. In particular, a reduction of the structure group for the tangent bundle provides important information about the manifold.

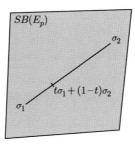

Figure 8.2. In the space of symmetric bilinear forms, all the points on the straight line between two positive definite forms are positive definite.

The bundle is *orientable*, see Section 6.7, if it is possible to choose an atlas whose transition functions map to the special linear group $SL_n(\mathbf{R})$. If we can reduce to the subgroup consisting of only the identity matrix, then the bundle is trivial. If $n = 2m$, then $GL_m(\mathbf{C}) \subseteq GL_n(\mathbf{R})$, and a reduction to $GL_m(\mathbf{C})$ is called a *complex structure* on the bundle.

Here is an example of the unreasonable power of the existence of fiber metrics, making the theory of smooth vector bundles much easier than some of its neighboring fields.

Proposition 8.3.4   *Surjective morphisms of smooth bundles split, i.e., if*

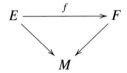

is a surjective smooth bundle morphism, then there is a bundle morphism $s \colon F \to E$ over M such that $fs = \mathrm{id}_F$.

Proof.   Assume E has rank n and F has rank k. By the surjectivity assumption f has constant rank k and Corollary 6.1.10 gives that $K = \ker\{f\} \subseteq E$ is a subbundle of rank $n - k$. Since F and E/K have equal rank, the surjective bundle morphism $E/K \to F$ defined by $\bar{v} \mapsto f(v)$ is an isomorphism. Choosing a fiber metric, we get from Lemma 6.6.13 that $K^\perp \subseteq E \to E/K$ is an isomorphism. Let s be the resulting composite

$$F \xleftarrow{\;\cong\;} E/K \xleftarrow{\;\cong\;} K^\perp \subseteq E.$$

By construction the composite $fs \colon F \to F$ is the identity.   ■

As further evidence of the simplifications made possible by the combination of Theorem 8.3.1 and Lemma 6.6.13 we offer the following string of exercises, culminating in the tantalizing isomorphism $T\mathbf{RP}^n \oplus \epsilon \cong \eta_n \oplus \cdots \oplus \eta_n$.

Recall from Definition 6.5.1 that, if $N \subseteq M$ is a smooth submanifold, the normal bundle $\perp N \to N$ is defined as the quotient bundle $(TM|_N)/TN \to N$, which – subject to a choice of Riemannian metric – is isomorphic to $(TN)^\perp \to N$.

Exercise 8.3.5   Let $M \subseteq \mathbf{R}^n$ be a smooth submanifold. Prove that $\perp M \oplus TM \to M$ is trivial.

Exercise 8.3.6   Consider S^n as a smooth submanifold of $\mathbf{R}^{n+1}$ in the usual way. Prove that the normal bundle is trivial.

Exercise 8.3.7

The tautological line bundle $\eta_n \to \mathbf{RP}^n$ is a subbundle of the trivial bundle $\mathrm{pr} \colon \mathbf{RP}^n \times \mathbf{R}^{n+1} \to \mathbf{RP}^n$:

$$\eta_n = \{(L, v) \in \mathbf{RP}^n \times \mathbf{R}^{n+1} \mid v \in L\} \subseteq \mathbf{RP}^n \times \mathbf{R}^{n+1} = \epsilon.$$

Let

$$\eta_n^\perp = \{\{(L, v) \in \mathbf{RP}^n \times \mathbf{R}^{n+1} \mid v \in L^\perp\} \subseteq \mathbf{RP}^n \times \mathbf{R}^{n+1} = \epsilon$$

be the orthogonal complement.

Prove that the Hom-bundle $\mathrm{Hom}(\eta_n, \eta_n^\perp) \to \mathbf{RP}^n$ is isomorphic to the tangent bundle $T\mathbf{RP}^n \to \mathbf{RP}^n$.

Exercise 8.3.8

Let $\epsilon = \mathbf{RP}^n \times \mathbf{R} \to \mathbf{RP}^n$ be the product line bundle. Prove that there is an isomorphism of bundles between $T\mathbf{RP}^n \oplus \epsilon$ and the $(n + 1)$-fold sum of the tautological line bundle $\eta_n \to \mathbf{RP}^n$ with itself:

$$T\mathbf{RP}^n \oplus \epsilon \cong \eta_n \oplus \cdots \oplus \eta_n.$$

Prove that "$4\eta_3$" and "$4\eta_2$" are trivial.

Exercise 8.3.9

Do this exercise only if you know about rings and modules. If $E \to M$ is a smooth vector bundle, let $\Gamma(E)$ be the vector space of smooth sections. Give $\Gamma(E)$ a $\mathcal{C}^\infty(M)$-module structure. Assume that $E \to M$ is a subbundle of a trivial bundle. Prove that $\Gamma(E)$ is a projective $\mathcal{C}^\infty(M)$-module.

Also, give the set of smooth bundle morphisms between two vector bundles over M a $\mathcal{C}^\infty(M)$-module structure such that the one-to-one correspondence between bundle morphisms and sections in the Hom-bundle of Exercise 6.4.4 becomes a $\mathcal{C}^\infty(M)$-isomorphism.

8.4 An Introduction to Morse Theory

Morse theory (named after its inventor Marston Morse, 1892–1977)[2] essentially states that the critical points of any sufficiently nice smooth function $f \colon M \to \mathbf{R}$ contain all the information necessary to build the manifold M. Hence it is an eminent example of how local information (critical points) occasionally can be gathered to give a global picture.

This is important both for practical and for theoretical reasons. As we shall see in Theorem 8.4.16, smooth functions are sometimes able to tell when a manifold is homeomorphic to a sphere, which was how Milnor could know that what turned out to be an exotic sphere actually was homeomorphic to a sphere. In applied topology, ideas from Morse theory are used to recognize the shape of complex data. The list showing the usefulness of Morse theory could go on, spanning from quantum field theory to Lie groups.

[2] https://en.wikipedia.org/wiki/Marston_Morse

Raoul Bott (1923–2005)[3] used Morse theory to show the periodicity theorem. Periodicity implies, among other things, the amazing fact that (modulo stably trivial bundles) vector bundles over S^n depend only on n modulo 8 – so our classification of vector bundles over the circle in Exercise 5.3.15 is relevant to S^9, S^{17} ... as well!

Proving periodicity is beyond our scope, but the reader may enjoy consulting, e.g., Milnor's book [14] for this and much else relating to Morse theory. Most of the material we *will* cover requires no heavy machinery, but a crucial point (Proposition 8.4.14) uses both the existence of Riemannian metrics and the integrability theorem.

Example 8.4.1   Imagine you are a fish near the shore, but can only see things under water. A standing torus is placed on the beach and the tide is coming in. Initially you see only the base of the torus. Shortly it grows into a bowl with the surface of the water touching the rim in a nice circle, and perhaps you imagine that this is the base of a sphere (both tori and spheres are fairly commonplace at the beach).

This remains the state of affairs for a while, but all of a sudden something dramatic happens: the water reaches the hole in the torus and the surface of the water touches the torus in something that looks like a figure eight.

This is the only dramatic event for a while: for some time the surface of the water touches the torus in two disjoint circles – like in the top picture in Figure 8.3.

Two more noteworthy incidents occur: when the two circles come together and finally when the top is capped off at the moment the torus gets fully submerged – a situation between these two events is pictured at the bottom in Figure 8.3.

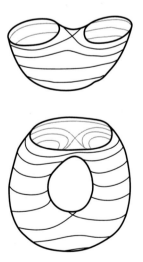

Figure 8.3.

[3] https://en.wikipedia.org/wiki/Raoul_Bott

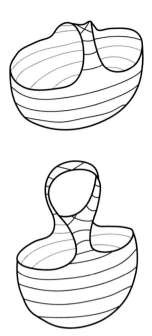

Figure 8.4.

The fish (and homotopy theorists) would say that the torus can be described by the following procedure (Figure 8.4).

1. Start with a point (which grows into a disk).
2. Attach a handle to get a basket.
3. Attach a new handle on top of the old one so that the rim is again a circle.
4. Attach a disk to this circle.

The point is that all this information on how to assemble a torus by "attaching handles" is hidden in a single function given by the height; and the places where anything interesting happens are exactly the critical points of this function.

This way of viewing a torus isn't actually all that fishy. Consider the usual flat representation of the torus obtained from identifying opposite edges of a square. Start with a small disk D around the common corner point. Fatten up the two edges to ribbons A and B which we attach to the small disk. Finally, attach a disk S to fill the square. See Figures 8.5 and 8.6.

Note 8.4.2   *How* to attach handles is determined by something called the "index" of the critical point, which is calculated concretely in terms of second derivatives. For functions $f : \mathbf{R} \to \mathbf{R}$ this is well known from high school. The critical points are the points p where $f'(p) = 0$. If $f''(p) > 0$ (resp. $f''(p) < 0$) then p is a local minimum (resp. maximum) of the graph. One dimension up, it gets only a little more complicated. If $f : \mathbf{R}^2 \to \mathbf{R}$ the critical points are the points p where

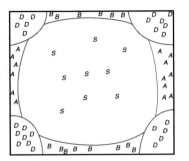

Figure 8.5. The torus assembled from a disk D, two ribbons A and B and disk S.

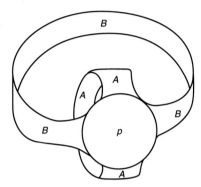

Figure 8.6. A view of D, A and B with identifications performed – do you recognize the basket with two handles?

the Jacobian vanishes $Df(p) = 0$. The nature of the critical point is given by the Hessian matrix

$$\begin{bmatrix} D_1 D_1 f(p) & D_1 D_2 f(p) \\ D_2 D_1 f(p) & D_2 D_2 f(p) \end{bmatrix}.$$

If it has two negative (resp. positive) eigenvalues p is a local minimum (resp. maximum) and if it has one positive and one negative eigenvalue it is a saddle point.

From the point of view of f being the height function of its graph, a minimum is like starting with a disk and a maximum is like capping off with one. A saddle point is building a handle.

Note 8.4.3 A *critical* point of a smooth map $f: M \to \mathbf{R}$ is a point $p \in M$ where the differential $d_p f \in T_p^* M$ is zero, which is the same as saying that the Jacobian (in some chart) is zero. This agrees with Definition 4.1.8, where a critical point was said to be a point $p \in M$ such that $T_p f: T_p M \to T_{f(p)} \mathbf{R} \cong \mathbf{R}$ has rank zero (c.f. also Exercise 4.1.12).

Definition 8.4.4 We say that a critical point $p \in M$ of a smooth map $f : M \to \mathbf{R}$ is *non-degenerate* if df and the zero section are transverse at p. We say that f is a *Morse function* if df is transverse to the zero section, i.e., if all critical points are non-degenerate.

Lemma 8.4.5 *The critical points of a Morse function are isolated.*

Proof. We can assume that $M = \mathbf{R}^n$. Then the Morse condition states that the composite

$$\mathbf{R}^n \to T^*\mathbf{R}^n \cong \mathbf{R}^n \times \mathbf{R}^n, \quad p \mapsto (p, [Df(p)]^T)$$

is transverse to $\mathbf{R}^n \times \{0\}$, i.e., we have that $[Df(q)]^T = 0$ implies that the Jacobian of $p \mapsto [Df(p)]^T$ at q is nonzero. The inverse function theorem, 4.2.1, applied to $p \mapsto [Df(p)]^T$ then implies that there is a neighborhood around q such that $p \mapsto [Df(p)]^T$ is a diffeomorphism, so that q is the only critical point in this neighborhood. ∎

We noticed in the proof that (in some chart) the "Jacobian of the Jacobian" played a rôle. This symmetric matrix of all second partial derivatives is important enough to have its own name: the Hessian matrix.

Definition 8.4.6 Let $f : M \to \mathbf{R}$ be smooth with critical point p. The *Hessian matrix* of f at p with respect to a chart (x, U) around p is the matrix

$$D(D(fx^{-1}))(x(p)).$$

The *index* of f at p is the sum of the dimensions of the eigenspaces of the Hessian with negative eigenvalues.

Again, the rank and index of the Hessian are independent of the choice of chart (check this!)

As in the proof of Lemma 8.4.5, the transversality of df and the zero section at p is equivalent to the Hessian matrix being invertible, giving the following lemma.

Lemma 8.4.7 *A critical point p of a smooth function $f : M \to \mathbf{R}$ is non-degenerate if and only if (in some chart) the Hessian matrix has maximal rank.*

Example 8.4.8 Consider the critical point 0 of the function $f : \mathbf{R} \to \mathbf{R}$, $f(t) = t^3$. The Jacobian (in the identity chart) is $f'(t) = 3t^2$ and the Hessian is $f''(t) = 6t$. Since $f''(0) = 0$ the critical point is not non-degenerate (aka degenerate).

Note that adding an arbitrarily small quadratic term (dampened by an appropriate bump function) will change f into a Morse function. This is an example of a general phenomenon: up to an arbitrarily small correction term, any smooth function is a Morse function.

Exercise 8.4.9   Continuing the robot example, show that the length $f: S^1 \times S^1 \to \mathbf{R}^1$, $f(e^{i\theta}, e^{i\phi}) = \sqrt{11 - 6\cos\theta - 6\cos\phi + 2\cos(\theta - \phi)}$ of the telescopic arm is a Morse function and calculate the index at each critical point.

Exercise 8.4.10   Let $a \in \mathbf{R}$ and consider the smooth function $f: \mathbf{R} \to \mathbf{R}$, $f(t) = t^4 - at^2$. For what a is f Morse?

We have the following extension of the rank theorem to non-degenerate critical points. If p is critical the linear part of f near p is of course zero, but the quadratic part takes a very special form.

Lemma 8.4.11 (The Morse Lemma)   *Let p be a non-degenerate critical point of index k of the smooth map $f: M \to \mathbf{R}$. Then there is a chart (x, U) around p such that for $t = (t_1, \dots, t_n) \in x(U)$ we have that*

$$fx^{-1}(t) = -\sum_{i=1}^{k} t_i^2 + \sum_{i=k+1}^{n} t_i^2.$$

Proof.   On choosing an arbitrary chart we are reduced to the case where $(M, p) = (\mathbf{R}^n, 0)$. Now, recall that Lemma 3.4.8 showed that

$$f(t) = \sum_{i=1}^{n} t_i f_i(t),$$

where $f_i(0) = D_i f(0) = 0$. Applying Lemma 3.4.8 to each of the f_i we get that

$$f(t) = \sum_{i,j} t_i t_j f_{ij}(t) = \sum_{i,j} t_i t_j h_{ij}(t), \qquad \text{where } h_{ij}(t) = \frac{f_{ij}(t) + f_{ji}(t)}{2}.$$

Now, $h_{ij}(t)$ is symmetric, so we can orthogonally diagonalize it, the only possible problem being that this must be done smoothly with respect to t. Furthermore, on taking all the second partial differentials we see that the matrix $[h_{ij}(0)]$ actually is the Hessian of f at 0, and hence it is non-degenerate. Consequently, in a neighborhood of 0 the eigenvalues will all be different from zero. By a change of coordinates, we may assume $h_{11}(t) \neq 0$ in a neighborhood.

Let $x(t) = (x_1(t), \dots, x_n(t))$ where $x_1(t) = \sqrt{|h_{11}(t)|}\left(t_1 + \sum_{i=2}^{n}(h_{i1}(t)/h_{11}(t))t_i\right)$ and $x_j(t) = t_j$ for $j > 1$. Since the Jacobian $Dx(0)$ is invertible ($\det Dx(0) = \sqrt{|h_{11}(0)|}$) there is a neighborhood of 0 where x defines a smooth chart. Upon inserting

$$t_1 = \frac{x_1(t)}{\sqrt{|h_{11}(t)|}} - \sum_{j=2}^{n} \frac{h_{i1}(t)}{h_{11}(t)} t_i$$

into our expression for f we get that

$$fx^{-1}(s) = \frac{h_{11}x^{-1}(s)}{|h_{11}x^{-1}(s)|}s_1^2 + \sum_{i,j>1} s_i s_j \tilde{h}_{ij}(s)$$

with $[\tilde{h}_{ij}(s)]$ symmetric and $[\tilde{h}_{ij}(0)]$ invertible. By iterating this process n times we obtain a sum of squares with coefficients ± 1. The index of a symmetric matrix is not changed upon orthogonal diagonalization, so by rearranging the summands we get the desired form. ∎

Note 8.4.12   Morse functions are everywhere: given any smooth $f : M \to \mathbf{R}$ and $\epsilon > 0$ there is a Morse function g such that for all $p \in M$ $|f(p) - g(p)| < \epsilon$ (see, e.g., Section IV3.4 of [11]).

Let M be a compact manifold, $f : M \to \mathbf{R}$ a smooth function and a a regular value. Then $f^{-1}(a) \subseteq M$ is a smooth submanifold of codimension 1 (or empty), $f^{-1}((-\infty, a)) \subseteq M$ is an open submanifold and

$$M^a = f^{-1}((-\infty, a])$$

is a manifold with boundary $f^{-1}(a)$. We will state most results in terms of M^a, but if you are uncomfortable with manifolds with boundary, much can be gained by looking at $f^{-1}((-\infty, a))$ instead.

Exercise 8.4.13   Check that, when a is regular, $M^a = f^{-1}((-\infty, a])$ is indeed a manifold with boundary.

The following proposition gives substance to the notion that "nothing happens between critical points". The compactness condition is not important, it just allows us to refer to the compact case of the integrability theorem, 7.2.1, in order to avoid giving a separate argument proving that a certain flow is global.

Proposition 8.4.14   *Let M be a compact manifold, $f : M \to \mathbf{R}$ a smooth function and $a < b \in \mathbf{R}$. Suppose f has no critical values in $[a, b]$. Then M^a is diffeomorphic to M^b.*

Proof.   Let $\sigma : M \to \mathbf{R}$ be a bump function with $f^{-1}([a, b]) \subseteq \sigma^{-1}(1)$ and such that the support of σ contains no critical points. Choose a Riemannian metric g on M. Recall that the gradient $\operatorname{grad} f : M \to TM$ is defined in Definition 6.6.8 through the equation $df = g(\operatorname{grad} f, -)$. Let $X : M \to TM$ be the vector field defined by

$$X(p) = \begin{cases} \dfrac{\sigma(p)}{g(\operatorname{grad}_p f, \operatorname{grad}_p f)} \operatorname{grad}_p f & p \in \operatorname{supp}(\sigma) \\ 0 & \text{otherwise.} \end{cases}$$

Let Φ be the global flow corresponding to X provided by the integrability theorem, 7.2.1. The associated diffeomorphism $\Phi_{b-a}\colon M \cong M$ of Note 7.1.12 will take M^a diffeomorphically onto its image, and our task is then to show that this image is precisely M^b.

Let ϕ_p be the flow line through $p \in M$. Observe that

$$(f\phi_p)'(0) = df([\phi_p]) = df(X(p))$$
$$= g(\mathrm{grad}_p f, X(p))$$
$$= \sigma(p).$$

Hence, if $\Phi(s, p) \in \sigma^{-1}(1)$ for s between 0 and t we have that

$$f\Phi(t, p) = t + f(p),$$

with slower movement when we stray outside $\sigma^{-1}(1)$ (even constant outside $\mathrm{supp}\,\sigma$).

Consequently, if $p \in M^a$, then $f(\Phi_{b-a}(p)) \leq b - a + f(p) \leq b - a + a = b$ and if $p \in M^b$, then $p = \Phi_{b-a}(\Phi_{a-b}(p))$ and $f(\Phi_{a-b}(p)) \leq a$. See Figure 8.7. ∎

Note 8.4.15 The proof of Proposition 8.4.14 actually provides us with what is called a *deformation retraction* of M^a in M^b (which is stronger than saying that the inclusion $M^a \subseteq M^b$ is a "homotopy equivalence"). Roughly it says that we can

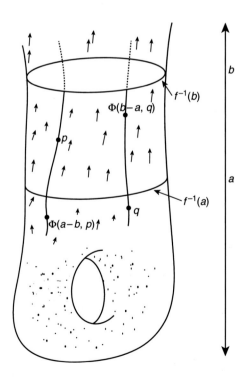

Figure 8.7.

deform M^a to gradually get M^b. More precisely we have the following. In addition to the inclusion $M^a \subseteq M^b$ we have a (continuous) "homotopy"

$$H : [0, 1] \times M^b \to M^b$$

defined by

$$H(t, q) = \begin{cases} q & \text{if } f(q) \leq a \\ \Phi_{t(a - f(q))}(q) & \text{if } a \leq f(q) \leq b, \end{cases}$$

which has the property that for $q \in M^b$ we have $H(0, q) = q$ and $H(1, q) \in M^a$, whereas for $q \in M^a$ and *any* $t \in [0, 1]$ we have $H(t, q) = q$.

Here is a striking consequence, originally due to Georges Reeb (1920–1993)[4], which allows us to recognize a sphere when we see one – at least up to homeomorphism. This is exactly what made it possible for Milnor to show that the smooth manifolds he constructed (diffeomorphic to the Brieskorn manifolds discussed in Note 4.5.8) were homeomorphic to the 7-sphere.

Theorem 8.4.16 (Reeb) *A compact manifold supporting a Morse function with exactly two critical points is homeomorphic to a sphere.*

Proof.   Let $f : M \to \mathbf{R}$ be such a Morse function. The critical points must necessarily be the maximum and the minimum, for simplicity say $\min f = 0$ and $\max f = 1$. By the Morse lemma, 8.4.11, there is an $\epsilon > 0$ such that both M^ϵ and $f^{-1}[1 - \epsilon, 1]$ are diffeomorphic to disks. Since there are no critical values between 0 and 1, M^ϵ is diffeomorphic to $M^{1-\epsilon}$.

Hence, M is obtained by taking two spaces $M^{1-\epsilon}$ and $f^{-1}[1 - \epsilon, 1]$ – both of which are diffeomorphic to disks – and gluing them together along their common boundary – which is diffeomorphic to a sphere in two ways. Such a thing must necessarily be homeomorphic to a sphere, a fact which can be seen as follows.

Given a continuous $h : S^{n-1} \to S^{n-1}$, we extend it to a continuous $C_h : D^n \to D^n$ by $C_h(p) = |p| h(p/|p|)$ if $p \neq 0$ and $C_h(0) = 0$. If h is a homeomorphism, then so is C_h. Now, consider the situation where a space X is the union of X_1 and X_2 and we have homeomorphisms $g_i : D^n \to X_i$ that restrict to homeomorphisms $h_i : S^{n-1} \to X_1 \cap X_2$. Define the homeomorphism $S^n \to X$ via g_1 on the Northern Hemisphere and $g_2 C_{h_2^{-1} h_1}$ on the Southern Hemisphere. ∎

Note 8.4.17   The extension idea used in the proof of Theorem 8.4.16 doesn't work in the smooth setting. For instance, if $h : S^1 \to S^1$ is given by $h(z) = z^2$, then the proposed $C_h : D^2 \to D^2$ is not smooth at the origin (h is not a diffeomorphism, but illustrates the problem).

---

[4] https://en.wikipedia.org/wiki/Georges_Reeb

Note 8.4.18   Morse theory goes on from here to describe how manifolds may be obtained by "attaching handles". Unfortunately this will take us a bit too far afield from our main focus, so we will content ourselves with stating the first main result. The reader is encouraged to refer to Milnor [14] or Section VII of Kosinski [11].

Definition 8.4.19   Let $\phi\colon S^n \to X$ be continuous. Let $X \coprod_\phi D^{n+1}$ be the quotient space $X \coprod D^{n+1}/\sim$ under the equivalence relation that is $p \in S^n$, then $\phi(p) \in X$ is equivalent to $p \in D^{n+1}$. We say that $X \coprod_\phi D^{n+1}$ is obtained from X by *attaching an $(n + 1)$-cell.*

Example 8.4.20   Consider the usual flat representation of the torus obtained from identifying opposite edges of a square. The common corner is a 0-cell. The two edges are two 1-cells attached to the point. Finally, we attach a 2-cell to fill the square.

Theorem 8.4.21   *Let M be compact and let $f\colon M \to \mathbf{R}$ be a Morse function. Assume $f^{-1}[a, b]$ contains a single critical point p with $c = f(p) \in (a, b)$ and let λ be the index of f at p. Then M^b is obtained from M^a by attaching a λ-cell.*

Example 8.4.22   This is exactly what happened to our fish on the beach in Section 8.4, observing a torus gradually submerged as the tide came in. The height function underwent critical points of index 0, 1, 1 and 2 corresponding to the attaching of cells eventually resulting in the full torus.

Returning to the robot example, the length of the telescopic arm gave *another* Morse function with four critical points with indices – according to Exercise 8.4.9 – 0, 1, 1 and 2.

It should be noted that the cell decomposition is very sensitive to deformations of the Morse function.

8.5 Ehresmann's Fibration Theorem

We have studied quite intensively the consequences of a map $f\colon M \to N$ being an immersion. In fact, by adding the point set topological property that $M \to f(M)$ is a homeomorphism we got in Theorem 4.7.4 that f was an imbedding.

We are now ready to discuss submersions (i.e., smooth maps for which all points are regular). It turns out that by adding a point set property we get that submersions are also rather special: they look like vector bundles, except that the fibers are not vector spaces, but manifolds!

Definition 8.5.1 Let $f \colon E \to M$ be a smooth map. We say that f is a *locally trivial fibration* if for each $p \in M$ there is an open neighborhood U around p and a diffeomorphism $h \colon f^{-1}(U) \to U \times f^{-1}(p)$ such that

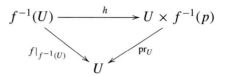

commutes.

Example 8.5.2 The projection of the torus down to a circle, which is illustrated in Figure 8.8, is kind of misleading since the torus is **globally** a product. However, due to the scarcity of compact two-dimensional manifolds, the torus is the only example of a total space of a locally trivial fibration with non-discrete fibers that can be imbedded in $\mathbf{R}^3$.

However, there are nontrivial examples we can envision: for instance, the projection of the Klein bottle onto its "central circle" (see Figure 8.9).

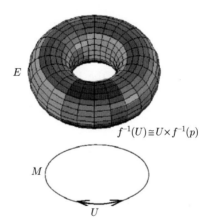

Figure 8.8. Over a small $U \in M$ a locally trivial fibration looks like the projection $U \times f^{-1}(p) \to U$

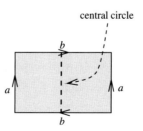

Figure 8.9. The projection from the Klein bottle onto its "central circle" is a locally trivial fibration.

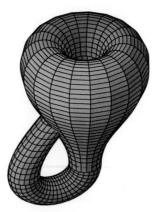

Figure 8.10. In the immersed picture of the Klein bottle, the "base space runs around the hole you see through". The fibers near the top are horizontal.

If we allow discrete fibers there are many examples small enough to be pictured. For instance, the squaring operation $z \mapsto z^2$ in complex numbers gives a locally trivial fibration $S^1 \to S^1$: the fiber of any point $z \in S^1$ is the set consisting of the two complex square roots of z (it is what is called a *double cover*). However, the fibration is not trivial (since S^1 is not homeomorphic to $S^1 \coprod S^1$)!

The last example is of a kind one encounters frequently: if $E \to M$ is a vector bundle endowed with some fiber metric, one can form the so-called *sphere bundle* $S(E) \to M$ by letting $S(E) = \{v \in E \mid |v| = 1\}$.

The double cover of S^1 above is exactly the sphere bundle associated with the infinite Möbius band, i.e., the line bundle with total space $[0, 1] \times \mathbf{R}/(0, t) \sim (1, -t)$. Similarly, the Klein bottle (Figure 8.10) is the sphere bundle associated with the vector bundle over S^1 with total space $[0, 1] \times \mathbf{R}^2/(0, (t_1, t_2)) \sim (1, (-t_1, t_2))$.

Recall from Exercise 5.2.5 that, up to isomorphism, there are exactly two vector bundles of a given rank r over S^1, and likewise there are exactly two r-sphere bundles over S^1: the trivial and the "generalized Klein bottle".

Exercise 8.5.3 Let $E \to M$ be a vector bundle. Show that $E \to M$ has a nonvanishing vector field if and only if the associated sphere bundle (with respect to some fiber metric) $S(E) \to M$ has a section.

Exercise 8.5.4 Prove that a smooth vector bundle with trivial sphere bundle is trivial. More precisely, if $\pi\colon E \to M$ is a smooth vector bundle of rank n such that the sphere bundle $S(\pi)\colon S(E) \to M$ with respect to some fiber metric is trivial (in the sense that there is a diffeomorphism $f\colon S(E) \overset{\cong}{\to} M \times S^{n-1}$ with $S(\pi) = \mathrm{pr}_M f$), then $\pi\colon E \to M$ is trivial.

Exercise 8.5.5 In a locally trivial (smooth) fibration over a connected smooth manifold all fibers are diffeomorphic.

The point set condition ensuring that submersions are locally trivial fibrations is the following.

Definition 8.5.6 A map $f \colon X \to Y$ is *proper* if the inverse images of compact subsets are compact.

Exercise 8.5.7 Prove that a locally trivial fibration with compact fibers is proper.

Exercise 8.5.8 Consider maps (and, as always, a "map" is continuous) $f \colon X \to Y$ and $g \colon Y \to Z$ of Hausdorff spaces. If f and g are proper then gf is proper. If gf is proper, then f is proper.

Exercise 8.5.9 The composite of two submersions is a submersion. If a composite gf is a submersion, then g is a submersion.

Theorem 8.5.10 (Ehresmann's Fibration Theorem) *Let $f \colon E \to M$ be a proper submersion. Then f is a locally trivial fibration.*

Proof. Since the question is local in M, we may start out by assuming that $M = \mathbf{R}^n$. The theorem then follows from Lemma 8.5.12 below. ∎

Note 8.5.11 Before we start with the proof, it is interesting to see what the ideas are.

By the rank theorem a submersion looks locally (in E and M) like a projection (Figure 8.11)

$$\mathbf{R}^{n+k} \to \mathbf{R}^n \times \{0\} \cong \mathbf{R}^n$$

and so locally all submersions are trivial fibrations. We will use flows to glue all these pieces together using partitions of unity (Figure 8.12).

The clue is then that a point $(t, q) \in \mathbf{R}^n \times f^{-1}(p)$ should correspond to what you get if you flow away from q, first a time t_1 in the first coordinate direction, then a time t_2 in the second and so on.

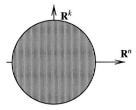

Figure 8.11. Locally a submersion looks like the projection from $\mathbf{R}^{n+k}$ down onto $\mathbf{R}^n$.

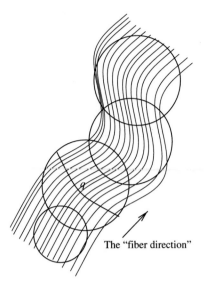

The "fiber direction"

Figure 8.12. The idea of the proof: make "flow" that flows transverse to the fibers: this is locally OK, but can we glue these pictures together?

Lemma 8.5.12 *Let* $f\colon E \to \mathbf{R}^n$ *be a proper submersion. Then there is a diffeomorphism* $h\colon E \to \mathbf{R}^n \times f^{-1}(0)$ *such that*

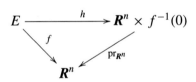

commutes.

Proof. If E is empty, the lemma holds vacuously since then $f^{-1}(0) = \emptyset$, and $\emptyset = \mathbf{R}^n \times \emptyset$. Disregarding this rather uninteresting case, let $p_0 \in E$ and $r_0 = f(p_0) \in \mathbf{R}^n$. The third part of the rank theorem, 4.3.3, guarantees that for all $p \in f^{-1}(r_0)$ there are charts $x_p\colon U_p \to U'_p$ such that

$$
\begin{array}{ccc}
U_p & \xrightarrow{\;f|_{U_p}\;} & \mathbf{R}^n \\[2pt]
{\scriptstyle x_p}\Big\downarrow & & \Big\| \\[2pt]
U'_p \subseteq \mathbf{R}^{n+k} & \xrightarrow{\;\text{pr}\;} & \mathbf{R}^n
\end{array}
$$

commutes (the map pr$\colon \mathbf{R}^{n+k} \to \mathbf{R}^n$ is the projection onto the first n coordinates).

Recall the "ith partial derivative"

$$
D_i\colon \mathbf{R}^m \to T\mathbf{R}^m, \qquad D_i(r) = [t \mapsto r + e_i t],
$$

where $e_i \in \mathbf{R}^m$ is the ith unit vector. Choose a partition of unity (see Theorem 8.2.2) $\{\phi_j\}$ subordinate to $\{U_p\}$. For every j choose a p such that supp$(\phi_j) \subseteq$

U_p, and let $x_j = x_p$. For $i = 1, \ldots, n$ we define the vector field (a "global ith partial derivative")

$$X_i \colon E \to TE, \qquad X_i(q) = \sum_j \phi_j(q) \cdot T(x_j^{-1}) D_i(x_j(q))$$

$$= \sum_j \phi_j(q) \cdot [t \mapsto x_j^{-1}(x_j(q) + e_i t)].$$

We claim that

$$
\begin{CD}
TE @>{Tf}>> T\mathbf{R}^n \\
@A{X_i}AA @AA{D_i}A \\
E @>>{f}> \mathbf{R}^n
\end{CD}
$$

commutes for all $i = 1, \ldots, n$. Indeed, using that $i \leq n$ and that $f|_{U_j} = \mathrm{pr}\, x_j$ we get that

$$T(\mathrm{pr}) D_i(x_j(u)) = D_i(\mathrm{pr}\, x_j(u)) = D_i(f(u))$$

for all $u \in U_j$, and so

$$Tf\, X_i(q) = \sum_j \phi_j(q) \cdot T(f x_j^{-1}) D_i(x_j(q))$$

$$= \sum_j \phi_j(q) \cdot T(\mathrm{pr}) D_i(x_j(q))$$

$$= \sum_j \phi_j(q) \cdot D_i(f(q))$$

$$= D_i(f(q)).$$

Fix the index i for a while. Notice that the curve $\beta \colon \mathbf{R} \to \mathbf{R}^n$ given by $\beta(t) = u + te_i$ is the unique solution to the initial value problem $\beta'(t) = e_i$, $\beta(0) = u$ (see Theorem 7.2.2), or, in terms of the velocity vector $\dot\beta \colon \mathbf{R} \to T\mathbf{R}^n$ given by $\dot\beta(t) = [s \mapsto \beta(s + t)]$ of Definition 7.1.13, β is unique with respect to the fact that $\dot\beta = D_i \beta$ and $\beta(0) = u$.

Let $\Phi_i \colon A_i \to E$ be the local flow corresponding to X_i, and let J_q be the slice of A_i at $q \in E$ (i.e., $A_i \cap (\mathbf{R} \times \{q\}) = J_q \times \{q\}$).

Fix q (and i), and consider the flow line $\alpha(t) = \Phi_i(t, q)$. Since flow lines are solution curves, the triangle in

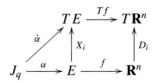

commutes, and since $Tf(\dot{\alpha}) = (\dot{f\alpha})$ and $(f\alpha)(0) = f(q)$ we get by uniqueness that

$$f\Phi_i(t, q) = f\alpha(t) = f(q) + te_i.$$

We want to show that $A_i = \mathbf{R} \times E$. Since $f\Phi_i(t, q) = f(q) + e_i t$ we see that the image of a bounded open interval under $f\Phi_i(-, q)$ must be contained in a compact set, say K. Hence the image of the bounded open interval under $\Phi_i(-, q)$ must be contained in $f^{-1}(K)$, which is compact since f is proper. But, if $J_q \neq \mathbf{R}$, then Lemma 7.3.6 tells us that $\Phi_i(-, q)$ will leave any given compact set in finite time, leading to a contradiction.

Hence all the Φ_i defined above are global and we define the diffeomorphism

$$\phi \colon \mathbf{R}^n \times f^{-1}(r_0) \to E$$

by

$$\phi(t, q) = \Phi_1(t_1, \Phi_2(t_2, \ldots, \Phi_n(t_n, q) \ldots)), \quad t = (t_1, \ldots, t_n) \in \mathbf{R}^n, \ q \in f^{-1}(r_0).$$

The inverse is given by

$$E \to \mathbf{R}^n \times f^{-1}(r_0),$$
$$q \mapsto (f(q) - r_0, \Phi_n((r_0)_n - f_n(q), \ldots, \Phi_1((r_0)_1 - f_1(q), q) \ldots)).$$

Finally, we note that we have also proven that f is surjective, and so we are free in our choice of $r_0 \in \mathbf{R}^n$. Choosing $r_0 = 0$ gives the formulation stated in the lemma. ∎

Corollary 8.5.13 (Ehresmann's Fibration Theorem, Compact Case)   *Let $f \colon E \to M$ be a submersion of compact smooth manifolds. Then f is a locally trivial fibration.*

Proof.   We need only notice that E being compact forces f to be proper: if $K \subset M$ is compact, it is closed (since M is Hausdorff), and $f^{-1}(K) \subseteq E$ is closed (since f is continuous). But a closed subset of a compact space is compact. ∎

Exercise 8.5.14   Check in all detail that the proposed formula for the inverse of ϕ given at the end of the proof of Ehresmann's fibration theorem, 8.5.10, is correct.

Exercise 8.5.15   Give an example of a submersion that is not a locally trivial fibration.

Exercise 8.5.16

Consider the projection

$$f\colon S^3 \to \mathbf{CP}^1.$$

Show that f is a locally trivial fibration. Consider the map

$$\ell\colon S^1 \to \mathbf{CP}^1$$

given by sending $z \in S^1 \subseteq \mathbf{C}$ to $[1, z]$. Show that ℓ is an imbedding. Use Ehresmann's fibration theorem to show that the inverse image

$$f^{-1}(\ell S^1)$$

is diffeomorphic to the torus $S^1 \times S^1$. (Note: there is a diffeomorphism $S^2 \to \mathbf{CP}^1$ given by $(a, z) \mapsto [1 + a, z]$, and the composite $S^3 \to S^2$ induced by f is called the *Hopf fibration* and has many important properties. Among other things it has the truly counter-intuitive property of detecting a "three-dimensional hole" in S^2!)

Exercise 8.5.17

Let $\gamma\colon \mathbf{R} \to M$ be a smooth curve and $f\colon E \to M$ a proper submersion. Let $p \in f^{-1}(\gamma(0))$. Show that there is a smooth curve $\sigma\colon \mathbf{R} \to E$ such that

commutes and $\sigma(0) = p$. Show that, if the dimensions of E and M agree, then σ is unique. In particular, study the cases and $S^n \to \mathbf{RP}^n$ and $S^{2n+1} \to \mathbf{CP}^n$.

Exercise 8.5.18

Consider the map

$$f\colon O(n+1) \to S^n, \qquad A \mapsto f(A) = Ae_1,$$

where $e_1 \in \mathbf{R}^{n+1}$ is the first standard basis vector. That is, $f(A)$ is the first column of A. Show that f is a locally trivial fibration with fiber diffeomorphic to $O(n)$.

Exercise 8.5.19

Prove that every smooth vector bundle over $\mathbf{R}^n$ is trivial.

Exercise 8.5.20

Let $\pi\colon E \to M$ be a locally trivial fibration and $f\colon N \to M$ a smooth map. Show that the induced map $\phi\colon E \times_M N \to N$ is a locally trivial fibration, where $E \times_M N$ is the fiber product defined in Exercise 4.7.11.

Exercise 8.5.21

Let $0 < m \le n$ and recall the *Milnor manifold* by

$$H(m, n) = \left\{ ([p], [q]) \in \mathbf{RP}^m \times \mathbf{RP}^n \mid \sum_{k=0}^{m} p_k q_k = 0 \right\},$$

which we proved in Exercise 2.5.18 was a smooth $(m + n - 1)$-dimensional submanifold of $\mathbf{RP}^m \times \mathbf{RP}^n$. Show that the projection $\pi\colon H(m, n) \to \mathbf{RP}^m$ sending $([p], [q])$ to $[p]$ is a locally trivial fibration with fiber $\mathbf{RP}^{n-1}$.

We end with three exercises giving (more) classical examples of something called *principal bundles* (which has to do with actions of Lie groups). There is a host of related exercises you can do, but I figured this would suffice.

Exercise 8.5.22   Let $0 < k < n$ and consider the map $f: \mathrm{SO}(n) \to V_n^k$ to the Stiefel manifold of Exercise 4.4.16 given by letting $f(A)$ be the $n \times k$ matrix consisting of the k first columns of A. Notice that this represents a k-frame: $f(A)^\mathrm{T} f(A)$ is the $k \times k$ identity matrix. Prove that f is a locally trivial fibration with all fibers diffeomorphic to $\mathrm{SO}(n-k)$.

Exercise 8.5.23   Let $0 < k < n$ and consider the map $C: \mathrm{SO}(n) \to \mathrm{Gr}(k, \mathbf{R}^n)$ to the Grassmann manifold of Example 2.3.15 given by sending a rotation matrix A to the subspace of $\mathbf{R}^n$ spanned by the first k columns of A: $C(A) = A \cdot \mathbf{R}^k \subseteq \mathbf{R}^n$ (we identify $\mathbf{R}^k$ with $\mathbf{R}^k \times \{0\} \subseteq \mathbf{R}^n$). Prove that C is a locally trivial fibration with all fibers diffeomorphic to $\mathrm{SO}(k) \times \mathrm{SO}(n-k)$.

Exercise 8.5.24   Let $0 < k < n$. Show that the map $S: V_n^k \to \mathrm{Gr}(k, \mathbf{R}^n)$ sending a frame to the subspace it spans is a locally trivial fibration with all fibers diffeomorphic to $\mathrm{SO}(k)$.

Appendix A
Point Set Topology

I have collected a few facts from point set topology. The aim is to present exactly what we need in the manifold course. Point set topology may be your first encounter of real mathematical abstraction, and can cause severe distress to the novice, but it is kind of macho when you get to know it a bit better. However, keep in mind that the course is about manifold theory, and point set topology is only a means of expressing some (obvious?) properties these manifolds should possess. Point set topology is a powerful tool when used correctly, but it is not our object of study.

The concept that typically causes most concern is the quotient space. This construction is used widely whenever we are working with manifolds and must be taken seriously. However, the abstraction involved should be eased by the many concrete examples (like the flat torus in the robot's arm in Example 1.1). For convenience I have repeated the definition of equivalence relations at the beginning of Section A.6.

If you need more details, consult any of the excellent books listed in the references. The real classics are [2] and [9], but the most widely used these days is [17]. There are also many on-line textbooks, some of which you may find at the Topology Atlas' "Education" website `http://at.yorku.ca/topology/educ.htm`.

Most of the exercises are not deep and are just rewritings of definitions (which may be hard enough if you are new to the subject), and the solutions are short.

If I list a fact without proof, the result **may** be deep and its proof (much too) hard.

At the end, or more precisely in Section A.10, I have included a few standard definitions and statements about sets that are used frequently both in the text and in the exercises. The purpose of collecting them in a section at the end is that, whereas they certainly should not occupy central ground in the text (or even in this appendix), the reader will still find the terms in the index and be referred directly to a definition, if she becomes uncertain about them at some point or other.

A.1 Topologies: Open and Closed Sets

· ·

> **Definition A.1.1**   Let X be a set. A *topology on X* is a family $\mathcal{U}$ of subsets
> of X with $\emptyset$, $X \in \mathcal{U}$ and which is closed under finite intersection and arbitrary
> unions, that is
>
> if $U, U' \in \mathcal{U}$, then $U \cap U' \in \mathcal{U}$,
> if $\mathcal{I} \subseteq \mathcal{U}$, then $\bigcup_{U \in \mathcal{I}} U \in \mathcal{U}$.
>
> We say that the pair $(X, \mathcal{U})$ is a *topological space*.

Frequently we will even refer to X as a topological space when $\mathcal{U}$ is evident from
the context.

Definition A.1.2   The members of $\mathcal{U}$ are called the *open sets* of X with respect to
the topology $\mathcal{U}$. A subset C of X is *closed* if the complement $X \setminus C = \{x \in X \mid x \notin C\}$ is open.

Exercise A.1.3   An open set on the real line $\mathbf{R}$ is a (possibly empty) union of open intervals. Check
that this defines a topology on $\mathbf{R}$. Check that the closed sets do **not** form a topology
on $\mathbf{R}$.

Definition A.1.4   A subset of X is called a *neighborhood* of $x \in X$ if it contains
an open set containing x.

Lemma A.1.5   *Let $(X, \mathcal{T})$ be a topological space. Prove that a subset $U \subseteq X$ is
open if and only if for all $p \in U$ there is an open set V such that $p \in V \subseteq U$.*

Proof.   Exercise!   ∎

Definition A.1.6   Let $(X, \mathcal{U})$ be a space and $A \subseteq X$ a subset. Then the *interior*
int A of A in X is the union of all open subsets of X contained in A. The *closure* $\bar{A}$
of A in X is the intersection of all closed subsets of X containing A.

Exercise A.1.7   Prove that int A is the biggest open set $U \in \mathcal{U}$ such that $U \subseteq A$, and that $\bar{A}$ is the
smallest closed set C in X such that $A \subseteq C$.

Exercise A.1.8   If (X, d) is a metric space (i.e., a set X and a symmetric positive definite function

$$d \colon X \times X \to \mathbf{R}$$

satisfying the triangle inequality), then X may be endowed with the *metric topology*
by letting the open sets be arbitrary unions of open balls (note: given an $x \in X$ and
a positive real number $\epsilon > 0$, the *open ϵ-ball centered in x* is the set $B(x, \epsilon) = \{y \in X \mid d(x, y) < \epsilon\}$). Exercise: show that this actually defines a topology.

In particular, *Euclidean n-space* is defined to be $\mathbf{R}^n$ with the metric topology.

Exercise A.1.9 The metric topology coincides with the topology we have already defined on **R**.

A.2 Continuous Maps

> **Definition A.2.1**   A *continuous map* (or simply a map)
> $$f : (X, \mathcal{U}) \to (Y, \mathcal{V})$$
> is a function $f : X \to Y$ such that for every $V \in \mathcal{V}$ the inverse image
> $$f^{-1}(V) = \{x \in X \mid f(x) \in V\}$$
> is in $\mathcal{U}$.

In other words, f is continuous if *the inverse images of open sets are open*.

Exercise A.2.2 Prove that a continuous map on the real line is just what you expect.
More generally, if X and Y are metric spaces, considered as topological spaces by giving them the metric topology as in Example A.1.8: show that a map $f : X \to Y$ is continuous if and only if the corresponding $\epsilon - \delta$-horror is satisfied.

Exercise A.2.3 Let $f : X \to Y$ and $g : Y \to Z$ be continuous maps. Prove that the composite $gf : X \to Z$ is continuous.

Example A.2.4   Let $f : \mathbf{R}^1 \to S^1$ be the map which sends $p \in \mathbf{R}^1$ to $e^{ip} = (\cos p, \sin p) \in S^1$. Since $S^1 \subseteq \mathbf{R}^2$, it is a metric space, and hence may be endowed with the metric topology. Show that f is continuous, and also that the images of open sets are open.

> **Definition A.2.5**   A *homeomorphism* is a continuous map $f : (X, \mathcal{U}) \to (Y, \mathcal{V})$ with a continuous inverse, that is a continuous map $g : (Y, \mathcal{V}) \to (X, \mathcal{U})$ with $f(g(y)) = y$ and $g(f(x)) = x$ for all $x \in X$ and $y \in Y$.

Exercise A.2.6 Prove that $\tan : (-\pi/2, \pi/2) \to \mathbf{R}$ is a homeomorphism.

Note A.2.7   Note that being a homeomorphism is **more than** being bijective and continuous. As an example, let X be the set of real numbers endowed with the metric topology, and let Y be the set of real numbers, but with the "indiscrete topology": only $\emptyset$ and Y are open. Then the identity map $X \to Y$ (sending the real number x to x) is continuous and bijective, but it is **not** a homeomorphism: the identity map $Y \to X$ is not continuous.

Definition A.2.8   We say that two spaces are *homeomorphic* if there exists a homeomorphism from one to the other.

A.3 Bases for Topologies

Definition A.3.1 If $(X, \mathcal{U})$ is a topological space, a subfamily $\mathcal{B} \subseteq \mathcal{U}$ is a *basis for the topology* $\mathcal{U}$ if for each $x \in X$ and each $V \in \mathcal{U}$ with $x \in V$ there is a $U \in \mathcal{B}$ such that

$$x \in U \subseteq V$$

as in Figure A.1.

Note A.3.2 This is equivalent to the condition that each member of $\mathcal{U}$ is a union of members of $\mathcal{B}$.

Conversely, given a family of sets $\mathcal{B}$ with the property that if $B_1, B_2 \in \mathcal{B}$ and $x \in B_1 \cap B_2$ then there is a $B_3 \in \mathcal{B}$ such that $x \in B_3 \subseteq B_1 \cap B_2$ (Figure A.2), we have that $\mathcal{B}$ is a basis for the topology on $X = \bigcup_{U \in \mathcal{B}} U$ given by declaring the open sets to be arbitrary unions from $\mathcal{B}$. We say that the basis $\mathcal{B}$ *generates* the topology on X.

Exercise A.3.3 The real line has a countable basis for its topology.

Exercise A.3.4 The balls with rational radius and whose centers have coordinates that all are rational form a countable basis for $\mathbf{R}^n$.

Just to be absolutely clear: a topological space $(X, \mathcal{U})$ has a *countable basis* for its topology if and only if there exists a countable subset $\mathcal{B} \subseteq \mathcal{U}$ which is a basis.

Figure A.1.

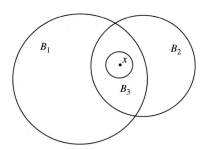

Figure A.2.

Exercise A.3.5
Let (X, d) be a metric space. Then the open balls form a basis for the metric topology.

Exercise A.3.6
Let X and Y be topological spaces, and $\mathcal{B}$ a basis for the topology on Y. Show that a function $f : X \to Y$ is continuous if $f^{-1}(V) \subseteq X$ is open for all $V \in \mathcal{B}$.

A.4 Separation

There are zillions of separation conditions, but we will be concerned only with the most intuitive of all: Hausdorff spaces (named after Felix Hausdorff (1868–1942))[1].

Definition A.4.1 A topological space $(X, \mathcal{U})$ is *Hausdorff* if for any two distinct $x, y \in X$ there exist disjoint neighborhoods of x and y (Figure A.3).

Example A.4.2 The real line is Hausdorff.

Example A.4.3 More generally, the metric topology is always Hausdorff.

A.5 Subspaces

Definition A.5.1 Let $(X, \mathcal{U})$ be a topological space. A *subspace* of $(X, \mathcal{U})$ is a subset $A \subset X$ with the topology given by letting the open sets be $\{A \cap U \mid U \in \mathcal{U}\}$ (Figure A.4).

Exercise A.5.2
Show that the subspace topology **is** a topology.

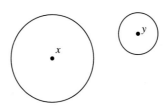

Figure A.3. The two points x and y are contained in disjoint open sets.

[1] https://en.wikipedia.org/wiki/Felix_Hausdorff

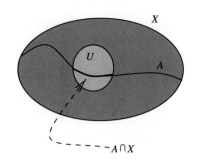

Figure A.4.

Prove that a map to a subspace $Z \to A$ is continuous if and only if the composite

$$Z \to A \subseteq X$$

is continuous.

Prove that, if X has a countable basis for its topology, then so has A.

Prove that, if X is Hausdorff, then so is A.

Corollary A.5.6 *All subspaces of $\mathbf{R}^n$ are Hausdorff, and have countable bases for their topologies.*

Definition A.5.7 If $A \subseteq X$ is a subspace, and $f \colon X \to Y$ is a map, then the composite

$$A \subseteq X \to Y$$

is called the *restriction of f to A*, and is written $f|_A$.

A.6 Quotient Spaces

Before defining the quotient topology we recall the concept of equivalence relations.

Definition A.6.1 Let X be a set. An *equivalence relation* on X is a subset E of the set $X \times X = \{(x_1, x_2) \mid x_1, x_2 \in X\}$ satisfying the following three conditions:

(reflexivity)	$(x, x) \in E$ for all $x \in X$,
(symmetry)	if $(x_1, x_2) \in E$ then $(x_2, x_1) \in E$,
(transitivity)	if $(x_1, x_2) \in E$ and $(x_2, x_3) \in E$, then $(x_1, x_3) \in E$.

We often write $x_1 \sim x_2$ instead of $(x_1, x_2) \in E$.

Definition A.6.2 Given an equivalence relation E on a set X we may for each $x \in X$ define the *equivalence class* of x to be the subset $[x] = \{y \in X \mid x \sim y\}$.

This divides X into a collection of nonempty, mutually disjoint subsets.
The set of equivalence classes is written $X/\sim$, and we have a surjective function

$$X \to X/\sim$$

sending $x \in X$ to its equivalence class $[x]$.

Definition A.6.3 Let $(X, \mathcal{U})$ be a topological space, and consider an equivalence relation $\sim$ on X. The *quotient space* with respect to the equivalence relation is the set $X/\sim$ with the *quotient topology*. The quotient topology is defined as follows. Let

$$p \colon X \to X/\sim$$

be the projection sending an element to its equivalence class. A subset $V \subseteq X/\sim$ is open if and only if $p^{-1}(V) \subseteq X$ is open (Figure A.5).

Exercise A.6.4 Show that the quotient topology **is** a topology on $X/\sim$.

Exercise A.6.5 Prove that a map from a quotient space $(X/\sim) \to Y$ is continuous if and only if the composite

$$X \to (X/\sim) \to Y$$

is continuous.

Exercise A.6.6 The projection $\mathbf{R}^1 \to S^1$ given by $p \mapsto e^{ip}$ shows that we may view S^1 as the set of equivalence classes of real number under the equivalence $p \sim q$ if there is an integer n such that $p = q + 2\pi n$. Show that the quotient topology on S^1 is the same as the subspace topology you get by viewing S^1 as a subspace of $\mathbf{R}^2$.

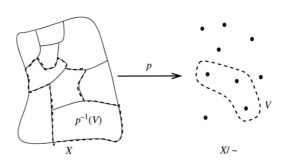

Figure A.5.

A.7 Compact Spaces

> **Definition A.7.1** A *compact space* is a space $(X, \mathcal{U})$ with the following property: in any set $\mathcal{V}$ of open sets covering X (i.e., $\mathcal{V} \subseteq \mathcal{U}$ and $\bigcup_{V \in \mathcal{V}} V = X$) there is a finite subset that also covers X.

Exercise A.7.2 If $f: X \to Y$ is a continuous map and X is compact, then $f(X)$ is compact.

We list without proof the following results. Remember that a subset $A \subseteq \mathbf{R}^n$ is *bounded* if the set of distances between points, $\{|p - q| \mid p, q \in A\} \subseteq \mathbf{R}$, is bounded above.

Theorem A.7.3 (Heine–Borel) *A subset of $\mathbf{R}^n$ is compact if and only if it is closed and bounded.*

Example A.7.4 Hence the unit sphere $S^n = \{p \in \mathbf{R}^{n+1} \mid |p| = 1\}$ (with the subspace topology) is a compact space.

Exercise A.7.5 The *real projective space* $\mathbf{RP}^n$ is the quotient space $S^n/\sim$ under the equivalence relation $p \sim -p$ on the unit sphere S^n. Prove that $\mathbf{RP}^n$ is a compact Hausdorff space with a countable basis for its topology.

Theorem A.7.6 *If X is a compact space, then all closed subsets of X are compact spaces.*

Theorem A.7.7 *If X is a Hausdorff space and $C \subseteq X$ is a compact subspace, then $C \subseteq X$ is closed.*

A very important corollary of the above results is the following theorem.

Theorem A.7.8 *If $f: C \to X$ is a continuous map where C is compact and X is Hausdorff, then f is a homeomorphism if and only if it is bijective.*

Exercise A.7.9 Prove Theorem A.7.8 using the results preceding it.

Exercise A.7.10 Prove in three or fewer lines the standard fact that a continuous function $f: [a, b] \to \mathbf{R}$ has a maximum value.

A last theorem sums up some properties that are preserved under formation of quotient spaces (under favorable circumstances). It is not optimal, but will serve our needs. You can extract a proof from the more general statement given on p. 148 of [9].

Theorem A.7.11 *Let X be a compact space, and let $\sim$ be an equivalence relation on X. Let $p\colon X \to X/\!\sim$ be the projection and assume that, if $K \subseteq X$ is closed, then $p^{-1}p(K) \subseteq X$ is closed too.*

If X is Hausdorff, then so is $X/\!\sim$.

If X has a countable basis for its topology, then so has $X/\!\sim$.

A.8 Product Spaces

Definition A.8.1 If $(X, \mathcal{U})$ and $(Y, \mathcal{V})$ are two topological spaces, then their *product* is the set $X \times Y = \{(x, y) \mid x \in X, y \in Y\}$ with a basis for the topology given by products of open sets $U \times V$ with $U \in \mathcal{U}$ and $V \in \mathcal{V}$.

There are two projections, $\mathrm{pr}_X\colon X \times Y \to X$ and $\mathrm{pr}_Y\colon X \times Y \to Y$. They are clearly continuous.

Exercise A.8.2 A map $Z \to X \times Y$ is continuous if and only if both the composites with the projections

$$Z \to X \times Y \to X$$

and

$$Z \to X \times Y \to Y$$

are continuous.

Exercise A.8.3 Show that the metric topology on $\mathbf{R}^2$ is the same as the product topology on $\mathbf{R}^1 \times \mathbf{R}^1$, and, more generally, that the metric topology on $\mathbf{R}^n$ is the same as the product topology on $\mathbf{R}^1 \times \cdots \times \mathbf{R}^1$.

Exercise A.8.4 If X and Y have countable bases for their topologies, then so has $X \times Y$.

Exercise A.8.5 If X and Y are Hausdorff, then so is $X \times Y$.

A.9 Connected Spaces

Definition A.9.1 A space X is *connected* if the only subsets that are both open and closed are the empty set and the set X itself.

Exercise A.9.2 The natural generalization of the intermediate value theorem is "If $f : X \to Y$ is continuous and X connected, then $f(X)$ is connected". Prove this.

Note A.9.3 We say that a space X is *path connected* if for any $p_0, p_1 \in X$ there is a continuous $f : [0, 1] \to X$ such that $f(0) = p_0$ and $f(1) = p_1$.

The unit interval $[0, 1]$ is connected (as is any other interval, bounded or not), and so Exercise A.9.2 implies that, if $f : [0, 1] \to X$ is any continuous map, then the image is a connected subspace. Hence, if X is path connected then X is connected.

The maximal connected subsets of a space are called the *connected components*. The *path components* are the maximal path connected subspaces. Manifolds (being locally homeomorphic to Euclidean spaces) are path connected if and only if they are connected, so, for manifolds, the path components and the connected component coincide.

Definition A.9.4 Let $(X_1, \mathcal{U}_1)$ and $(X_2, \mathcal{U}_2)$ be topological spaces. The *disjoint union* $X_1 \coprod X_2$ is the union of disjoint copies of X_1 and X_2 (i.e., the set of pairs (k, x), where $k \in \{1, 2\}$ and $x \in X_k$), where an open set is a union of open sets in X and Y.

Exercise A.9.5 Show that the disjoint union of two nonempty spaces X_1 and X_2 is not connected.

Exercise A.9.6 A map $X_1 \coprod X_2 \to Z$ is continuous if and only if both the composites with the injections

$$X_1 \subseteq X_1 \coprod X_2 \to Z$$

and

$$X_2 \subseteq X_1 \coprod X_2 \to Z$$

are continuous.

A.10 Set-Theoretical Stuff

The only purpose of this section is to provide a handy reference for some standard results in elementary set theory.

Definition A.10.1 Let $A \subseteq X$ be a subset. The *complement* of A in X is the subset

$$X \setminus A = \{x \in X \mid x \notin A\}.$$

Definition A.10.2 Let $f : X \to Y$ be a function. We say that f is *injective* (or one-to-one) if $f(x_1) = f(x_2)$ implies that $x_1 = x_2$. We say that f is *surjective* (or onto) if for every $y \in Y$ there is an $x \in X$ such that $y = f(x)$. We say that f is *bijective* if it is both surjective and injective.

Definition A.10.3 Let $A \subseteq X$ be a subset and $f : X \to Y$ a function. The *image* of A under f is the set

$$f(A) = \{y \in Y \mid \text{there exists an } a \in A \text{ such that } y = f(a)\}.$$

The subset $f(X) \subseteq Y$ is simply called the image of f.

If $B \subseteq Y$ is a subset, then the *inverse image* (or *preimage*) of B under f is the set

$$f^{-1}(B) = \{x \in X \mid f(x) \in B\}.$$

The subset $f^{-1}(Y) \subseteq X$ is simply called the preimage of f.

Exercise A.10.4   Prove that $f(f^{-1}(B)) \subseteq B$ and $A \subseteq f^{-1}(f(A))$.

Exercise A.10.5   Prove that $f : X \to Y$ is surjective if and only if $f(X) = Y$ and injective if and only if for all $y \in Y$ $f^{-1}(\{y\})$ consists of at most a single element.

Definition A.10.6   Let X be a set and $\{A_i\}_{i \in I}$ be a family of subsets. Then the union is the subset

$$\bigcup_{i \in I} A_i = \{x \in X \mid \text{there is an } i \in I \text{ with } x \in A_i\}$$

and the intersection is the subset

$$\bigcap_{i \in I} A_i = \{x \in X \mid x \in A_i \text{ for all } i \in I\}.$$

Lemma A.10.7 (De Morgan's Formulae)   *Let X be a set and $\{A_i\}_{i \in I}$ be a family of subsets. Then*

$$X \setminus \bigcup_{i \in I} A_i = \bigcap_{i \in I}(X \setminus A_i),$$

$$X \setminus \bigcap_{i \in I} A_i = \bigcup_{i \in I}(X \setminus A_i).$$

Apology: the use of the term *family* is just phony: to us a family is nothing but a set (so a "family of sets" is nothing but a set of sets).

Exercise A.10.8   Let $B_1, B_2 \subseteq Y$ and $f : X \to Y$ be a function. Prove that

$$f^{-1}(B_1 \cap B_2) = f^{-1}(B_1) \cap f^{-1}(B_2),$$
$$f^{-1}(B_1 \cup B_2) = f^{-1}(B_1) \cup f^{-1}(B_2),$$
$$f^{-1}(Y \setminus B_1) = X \setminus f^{-1}(B_1).$$

If in addition $A_1, A_2 \subseteq X$ then

$$f(A_1 \cup A_2) = f(A_1) \cup f(A_2),$$
$$f(A_1 \cap A_2) \subseteq f(A_1) \cap f(A_2),$$
$$f(X) \setminus f(A_1) \subseteq f(X \setminus A_1),$$
$$B_1 \cap f(A_1) = f(f^{-1}(B_1) \cap A_1).$$

Appendix B
Hints or Solutions to the Exercises

Below you will find hints for all the exercises. Some are very short, and some are almost complete solutions. Ignore them if you possibly can, but, if you are really stuck, take a *quick* peek and see whether you can get some inspiration. You can seriously undermine your efforts if you make a habit of looking at the hints too early. To partially compensate: if you looked at the hint and believe you understood everything, close the book and see whether you can **write down** an argument without referring back to these pages.

Chapter 1

Exercise 1.5.2

Draw a hexagon with identifications so that it represents a handle attached to a Möbius band. Try your luck at cutting and pasting this figure into a (funny-looking) hexagon with identifications so that it represents three Möbius bands glued together (remember that the cuts may cross your identified edges).

Exercise 1.5.3

First, notice that any line through the origin intersects the unit sphere S^2 in two antipodal points, so that $\mathbf{RP}^2$ can be identified with $S^2/p \sim -p$. Since any point on the Southern Hemisphere is in the same class as one on the Northern Hemisphere we may disregard (in standard imperialistic fashion) all the points on the Southern Hemisphere, so that $\mathbf{RP}^2$ can be identified with the Northern Hemisphere with antipodal points on the equator identified. On smashing down the Northern Hemisphere onto a closed disk, we get that $\mathbf{RP}^2$ can be identified with a disk with antipodal points on the boundary circle identified. By pushing in the disk so that we get a rectangle we get the equivalent picture (disregard the lines in the interior of the rectangle for now) shown in Figure B.1.

The two dotted diagonal lines in Figure B.1 represent a circle. Cut $\mathbf{RP}^2$ along this circle, yielding a Möbius strip (Figure B.2) and two pieces (Figure B.3) that glue together to a disk (the pieces have been straightened out at the angles of the rectangle, and one of the pieces has to be reflected before it can be glued to the other to form a disk).

Exercise 1.5.4

Do an internet search (check for instance Wikipedia) to find the definition of the Euler characteristic. To calculate the Euler characteristic of surfaces you can simply use our flat representations as polygons, just remembering what points and edges really are identified.

Exercise 1.5.5

The beings could triangulate their universe, counting the number of vertices V, edges E and surfaces F in this triangulation (this can be done in finite time). The Euler characteristic $V - E + F$ uniquely determines the surface.

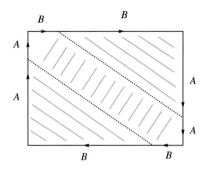

Figure B.1.

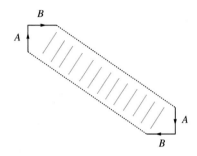

Figure B.2.

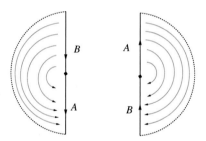

Figure B.3.

Chapter 2

Exercise 2.1.6

The map $x^{k,i}$ is the restriction of the corresponding projection $\mathbf{R}^{n+1} \to \mathbf{R}^n$ which is continuous, and the inverse is the restriction of the continuous map $\mathbf{R}^n \to \mathbf{R}^{n+1}$ sending $p = (p_0, \ldots, \widehat{p_k}, \ldots, p_n) \in \mathbf{R}^n$ (note the smart indexing) to $(p_0, \ldots, (-1)^i \sqrt{1 - |p|^2}, \ldots, p_n)$.

Exercise 2.1.7

(This exercise uses many results from Appendix A). Assume there was a chart covering all of S^n. That would imply that we had a homeomorphism $x \colon S^n \to U'$, where U' is an open subset of $\mathbf{R}^n$. For $n = 0$, this clearly is impossible since S^0 consists of two points, whereas $\mathbf{R}^0$ is a single point. Also for $n > 0$ this is impossible since S^n is compact (it is a bounded and closed subset of $\mathbf{R}^{n+1}$), and so $U' = x(S^n)$ would be compact (and nonempty), but $\mathbf{R}^n$ does not contain any compact **and** open nonempty subsets.

Exercise 2.2.8

Draw the lines in the picture in Example 2.2.7 and use high school mathematics to show that the formulae for $x^{\pm}$ are correct and define continuous functions (the formulae extend to functions defined on open sets in $\mathbf{R}^{n+1}$ where they are smooth and hence continuous). Then invert x^- and x^+, which again become continuous functions (so that $x^{\pm}$ are homeomorphisms), and check the chart transformation formulae.

Exercise 2.2.11

To get a smooth atlas, repeat the discussion in Example 2.1.8 for the real projective space, exchanging $\mathbf{R}$ with $\mathbf{C}$ everywhere. To see that $\mathbf{CP}^n$ is compact, it is convenient to notice that any $[p] \in \mathbf{CP}^n$ can be represented by $p/|p| \in S^{2n+1} \subseteq \mathbf{C}^n$, so that $\mathbf{CP}^n$ can alternatively be described as $S^{2n+1}/\sim$, where $p \sim q$ if there is a $z \in S^1$ such that $zp = q$. Showing that $\mathbf{CP}^n$ is Hausdorff and has a countable basis for its topology is not hard, but can be a bit more irritating, so a reference to Theorem A.7.11 provides an easy fix.

Exercise 2.2.12

All the chart transformations are identity maps, so the atlas is certainly smooth.

Exercise 2.2.13

Transport the structure radially out from the unit circle (i.e., use the homeomorphism from the unit circle to the square gotten by blowing up a balloon in a square box in flatland). All charts can then be taken to be the charts on the circle composed with this homeomorphism.

Exercise 2.2.14

The only problem is in the origin. If you calculate (one side of the limit needed in the definition of the derivative at the origin)

$$\lim_{t \to 0^+} \frac{\lambda(t) - \lambda(0)}{t} = \lim_{t \to 0^+} \frac{e^{-1/t}}{t} = \lim_{s \to \infty} \frac{s}{e^s} = 0,$$

you see that λ is once differentiable. It continues this way (you have to do a small induction showing that all derivatives at the origin involve limits of exponential times rational), proving that λ is smooth.

Exercise 2.3.4

If $\mathcal{B}$ is any smooth atlas containing $\mathcal{D}(\mathcal{A})$, then $\mathcal{D}(\mathcal{A}) \subseteq \mathcal{B} \subseteq \mathcal{D}(\mathcal{D}(\mathcal{A}))$. Prove that $\mathcal{D}(\mathcal{D}(\mathcal{A})) = \mathcal{D}(\mathcal{A})$.

Exercise 2.3.9

This is yet another of those "$\bar{x}\bar{y}^{-1} = (\bar{x}x^{-1})(xy^{-1})(y\bar{y}^{-1})$"-type arguments.

Exercise 2.3.10

It suffices to show that all the "mixed chart transformations" (like $x^{0,0}(x^+)^{-1}$) are smooth. Why?

Exercise 2.3.11

Because saying that "x is a diffeomorphism" is just a rephrasing of "$x = x(\text{id})^{-1}$ and $x^{-1} = (\text{id})x^{-1}$ are smooth". The charts in this structure are all diffeomorphisms $U \to U'$, where both U and U' are open subsets of $\mathbf{R}^n$.

Exercise 2.3.12

This will be discussed more closely in Lemma 8.1.1, but can be seen directly as follows. Since M is a topological manifold, it has a countable basis $\mathcal{B}$ for its topology. For each $(x, U) \in \mathcal{A}$ and $p \in U$ choose a $V \in \mathcal{B}$ with $p \in V \subseteq U$. The set of such sets V is countable. For each such V choose **one** of the charts $(x, U) \in \mathcal{A}$ with $V \subseteq U$.

Exercise 2.3.13   The mixed chart transformation consists of suitably restricted (co)sines and their inverses.

Exercise 2.3.16   Notice that x_V is a bijection, with inverse sending $f \in \mathrm{Hom}(V, V^\perp)$ to the graph $\Gamma(f) = \{v + f(v) \in \mathbf{R}^n \mid v \in V\} \subseteq \mathbf{R}^n$.

If $V, W \in \mathrm{Gr}(k, \mathbf{R}^n)$, then $x_V(U_V \cap U_W) = \{f \in \mathrm{Hom}(V, V^\perp) \mid \Gamma(f) \cap W^\perp = 0\}$. We must check that the chart transformation

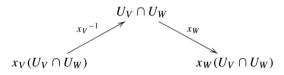

sending $f \colon V \to V^\perp$ to

$$W \xrightarrow{\ (\mathrm{pr}^W_{\Gamma(f)})^{-1}\ } \Gamma(f) \xrightarrow{\ \mathrm{pr}^{W^\perp}_{\Gamma(f)}\ } W^\perp$$

is smooth. For ease of notation we write

$$g_f = x_W x_V^{-1}(f) = \mathrm{pr}^{W^\perp}_{\Gamma(f)}(\mathrm{pr}^W_{\Gamma(f)})^{-1}$$

for this map (see Figure B.4). Now, if $x \in V$, then $(\mathrm{pr}^V_{\Gamma(f)})^{-1}(x) = x + f(x)$, and so the composite isomorphism

$$A_f = \mathrm{pr}^W_{\Gamma(f)}(\mathrm{pr}^V_{\Gamma(f)})^{-1} \colon V \to W$$

sending x to $A_f(x) = \mathrm{pr}^W x + \mathrm{pr}^W f(x)$ depends smoothly on f. By Cramer's rule, the inverse $B_f = A_f^{-1}$ also depends smoothly on f.

Finally, if $y \in W$, then

$$(\mathrm{pr}^W_{\Gamma(f)})^{-1}(y) = y + g_f(y)$$

is equal to

$$(\mathrm{pr}^V_{\Gamma(f)})^{-1}(B_f(y)) = B_f(y) + f(B_f(y)),$$

and so

$$g_f = B_f + f B_f - 1$$

depends smoothly on f.

The point set conditions are satisfied by the following purely linear algebraic assertions. For a subset $S \subseteq \{1, \dots, n\}$ of cardinality k, let $V_S \in \mathrm{Gr}(k, \mathbf{R}^n)$ be the subspace of all vectors $v \in \mathbf{R}^n$ with $v_j = 0$ for all $j \notin S$. The finite subcollection of $\mathcal{A}$ consisting of the U_{V_S} as S varies covers $\mathrm{Gr}(k, \mathbf{R}^n)$. If $W_1, W_2 \in \mathrm{Gr}(k, \mathbf{R}^n)$ there is a $V \in \mathrm{Gr}(k, \mathbf{R}^n)$ such that $W_1, W_2 \in U_V$. Explicitly, you may try something like this: decompose $W_1 = K \oplus K^\perp$ and $W_2 = L \oplus L^\perp$, where $K = \ker\{\mathrm{pr}^{W_1}_{W_2}\}$ and $L = \ker\{\mathrm{pr}^{W_2}_{W_1}\}$. Choosing as bases $(a_1, \dots, a_t)$ and $(b_1, \dots, b_t)$ for K and L, let $V = K^\perp \oplus M$, where M is spanned by $(a_1 + b_1, \dots, a_t + b_t)$ (check that this works).

Exercise 2.4.3   Given such charts, we prove that f is smooth at p. This implies that f is smooth since p is arbitrary.

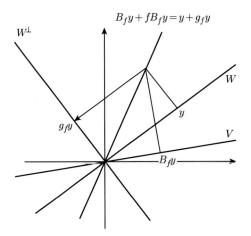

Figure B.4.

The function $f|_W$ is continuous since $y \circ f|_W \circ (x|_W)^{-1}$ is smooth (and hence continuous), and x and y are homeomorphisms. We must show that given **any** charts $(\tilde{x}, \tilde{U}) \in \mathcal{A}$ and $(\tilde{y}, \tilde{V}) \in \mathcal{B}$ with $p \in \tilde{W} = \tilde{U} \cap f^{-1}(\tilde{V})$ we have that $\tilde{y} f|_{\tilde{W}} (\tilde{x}|_{\tilde{W}})^{-1}$ is smooth at p. Now, for $q \in W \cap \tilde{W}$ we can rewrite the function in question as a composition

$$\tilde{y} f \tilde{x}^{-1}(q) = (\tilde{y} y^{-1})(y f x^{-1})(x \tilde{x}^{-1})(q)$$

of smooth functions defined on Euclidean spaces: $x \tilde{x}^{-1}$ and $y \tilde{y}^{-1}$ are smooth since $\mathcal{A}$ and $\mathcal{B}$ are smooth atlases.

Exercise 2.4.4

Use the identity chart on $\mathbf{R}$. The standard atlas on $S^1 \subseteq \mathbf{C}$ using projections is given simply by real and imaginary parts. Hence the formulae you have to check are smooth are sin and cos. This we know! One comment on domains of definition: let $f \colon \mathbf{R} \to S^1$ be the map in question; if we use the chart $(x^{0,0}, U^{0,0})$, then $f^{-1}(U^{0,0})$ is the union of all the intervals on the form $(-\pi/2 + 2\pi k, \pi/2 + 2\pi k)$ when k varies over the integers. Hence the function to check in this case is the function from this union to $(-1, 1)$ sending θ to $\sin \theta$.

Exercise 2.4.5

First check that $\tilde{g}$ is well defined ($g(p) = g(-p)$ for all $p \in S^2$). Check that it is smooth using the standard charts on $\mathbf{RP}^2$ and $\mathbf{R}^4$ (for instance: $\tilde{g} x^0(q_1, q_2) = (1/(1 + q_1^2 + q_2^2))(q_1 q_2, q_2, q_1, 1 + 2q_1^2 + 3q_2^2))$. To show that $\tilde{g}$ is injective, show that $g(p) = g(q)$ implies that $p = \pm q$.

Exercise 2.4.6

One way follows since the composite of smooth maps is smooth. The other follows since smoothness is a local question, and the projection $g \colon S^n \to \mathbf{RP}^n$ is a local diffeomorphism. More (or, perhaps, too) precisely, if $f \colon \mathbf{RP}^n \to M$ is a map, we have to show that, for all charts (y, V) on M, the composites $y f(x^k)^{-1}$ (defined on $U^k \cap f^{-1}(V)$) are smooth. But $x^k g(x^{k,0})^{-1} \colon D^n \to \mathbf{R}^n$ is a diffeomorphism (given by sending $p \in D^n$ to $(1/\sqrt{1 - |p|^2})p \in \mathbf{R}^n$), and so claiming that $y f(x^k)^{-1}$ is smooth is the same as claiming that $y(fg)(x^{k,0})^{-1} = y f(x^k)^{-1} x^k g(x^{k,0})^{-1}$ is smooth.

Exercise 2.4.11 Consider the map $f\colon \mathbf{RP}^1 \to S^1$ sending $[z]$ (with $z \in S^1 \subseteq \mathbf{C}$) to $z^2 \in S^1$, which is well defined since $(-z)^2 = z^2$. To see that f is smooth either consider the composite

$$S^1 \to \mathbf{RP}^1 \xrightarrow{f} S^1 \subseteq \mathbf{C}$$

(where the first map is the projection $z \mapsto [z]$) using Exercises 2.4.6 and 2.5.16, or do it from the definition: consider the standard atlas for $\mathbf{RP}^1$. In complex notation $U^0 = \{[z] \mid \mathrm{re}(z) \neq 0\}$ and $x^0([z]) = \mathrm{im}(z)/\mathrm{re}(z)$ with inverse $t \mapsto e^{i\tan^{-1}(t)}$. If $[z] \in U^0$, then $f([z]) = z^2 \in V = \{v \in S^1 \mid v \neq -1\}$. On V we choose the convenient chart $y\colon V \to (-\pi, \pi)$ with inverse $\theta \mapsto e^{i\theta}$, and notice that the "up, over and across" $yf(x^0)^{-1}(t) = 2\tan^{-1}(t)$ obviously is smooth. Likewise we cover the case $[z] \in U^1$. The method for showing that the inverse is smooth is similar.

Exercise 2.4.12 Consider the map $\mathbf{CP}^1 \to S^2 \subseteq \mathbf{R} \times \mathbf{C}$ sending $[z_0, z_1]$ to

$$\frac{1}{|z_0|^2 + |z_1|^2}\left(|z_1|^2 - |z_0|^2, 2z_1\overline{z_0}\right),$$

check that it is well defined and continuous (since the composite $\mathbf{C}^2 - \{0\} \to \mathbf{CP}^1 \to S^2 \subseteq \mathbf{R} \times \mathbf{C}$ is), calculate the inverse (which is continuous by Theorem A.7.8), and use the charts on S^2 from stereographic projection in Example 2.2.7 to check that the map and its inverse $(r, z) \mapsto [1 - r, z]$ are smooth. The reader may enjoy comparing the above with the discussion about qbits in Section 1.3.1.

Exercise 2.4.15 Given a chart (x, U) on M, define a chart $(xf^{-1}, f(U))$ on X.

Exercise 2.4.20 To see this, note that, given any p, there are open sets U_1 and V_1 with $p \in U_1$, $i(p) \in V_1$ and $U_1 \cap V_1 = \emptyset$ (since M is Hausdorff). Let $U = U_1 \cap i(V_1)$. Then U and $i(U) = i(U_1) \cap V_1$ do not intersect. As a matter of fact, M has a basis for its topology consisting of these kinds of open sets.

Carrying out shrinking even further, we may assume that U is a chart domain for a chart $x\colon U \to U'$ on M.

We see that $f|_U$ is open (it sends open sets to open sets, since the inverse image is the union of two open sets).

On U we see that f is injective, and so it induces a homeomorphism $f|_U\colon U \to f(U)$. We define the smooth structure on M/i by letting $x(f|_U)^{-1}$ be the charts for varying U. This is obviously a smooth structure, and f is a local diffeomorphism.

Exercise 2.4.21 Choose **any** chart $y\colon V \to V'$ with $p \in V$ in $\mathcal{U}$, then choose a small open ball $B \subseteq V'$ around $y(p)$. There exists a diffeomorphism h of this ball with all of $\mathbf{R}^n$. Let $U = y^{-1}(B)$ and define x by setting $x(q) = hy(q) - hy(p)$.

Exercise 2.5.4 Use "polar coordinates" (Figure B.5).

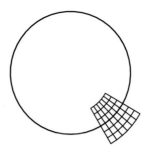

Figure B.5.

Exercise 2.5.5

Let $f(a_0, \ldots, a_{n-1}, t) = t^n + a_{n-1}t^{n-1} + \cdots + a_0$ and consider the map $x \colon \mathbf{R}^{n+1} \to \mathbf{R}^{n+1}$ given by

$$x(a_0, \ldots, a_n) = (a_1, \ldots, a_n, f(a_0, \ldots, a_n))$$

This is a smooth chart on $\mathbf{R}^{n+1}$ since x is a diffeomorphism with inverse given by sending $(b_1, \ldots, b_{n+1})$ to $(b_{n+1} - f(0, b_1, \ldots, b_n), b_1, \ldots, b_n)$. We see that $x(C) = \mathbf{R}^n \times 0$, and we have shown that C is an n-dimensional submanifold. Notice that we have only used that $f \colon \mathbf{R}^{n+1} \to \mathbf{R}$ is smooth and that $f(a_0, \ldots, a_n) = a_0 + f(0, a_1, a_n)$.

Exercise 2.5.6

Assume there is a chart $x \colon U \to U'$ with $(0, 0) \in U$, $x(0, 0) = (0, 0)$ and $x(K \cap U) = (\mathbf{R} \times 0) \cap U'$.

Then the composite (V is a sufficiently small neighborhood of 0)

$$V \xrightarrow{\;q \mapsto (q, 0)\;} U' \xrightarrow{\;x^{-1}\;} U$$

is smooth, and of the form $q \mapsto T(q) = (t(q), |t(q)|)$. But

$$T'(0) = \left(\lim_{h \to 0} \frac{t(h)}{h}, \lim_{h \to 0} \frac{|t(h)|}{h} \right),$$

and, for this to exist, we must have $t'(0) = 0$.

On the other hand, $x(p, |p|) = (s(p), 0)$, and we see that s and t are inverse functions. The directional derivative of $\mathrm{pr}_1 x$ at $(0, 0)$ in the direction $(1, 1)$ is equal to

$$\lim_{h \to 0^+} \frac{s(h)}{h},$$

but this limit does not exist since $t'(0) = 0$, and so x can't be smooth, which amounts to a contradiction.

Exercise 2.5.9

Let $f_1, f_2 \colon V \to \mathbf{R}^n$ be linear isomorphisms. Let G^1, G^2 be the two resulting smooth manifolds with underlying set $\mathrm{GL}(V)$. Showing that $G^1 = G^2$ amounts to showing that the composite

$$\mathrm{GL}(f_1)\mathrm{GL}(f_2)^{-1} \colon \mathrm{GL}(\mathbf{R}^n) \to \mathrm{GL}(\mathbf{R}^n)$$

is a diffeomorphism. Noting that $\mathrm{GL}(f_2)^{-1} = \mathrm{GL}(f_2^{-1})$ and $\mathrm{GL}(f_1)\mathrm{GL}(f_2^{-1}) = \mathrm{GL}(f_1 f_2^{-1})$, this amounts to showing that, given a fixed invertible matrix A

(representing $f_1 f_2^{-1}$ in the standard basis), conjugation by A, i.e., $B \mapsto ABA^{-1}$, is a smooth map $M_n(\mathbf{R}) \to M_n(\mathbf{R})$. This is true since addition and multiplication are smooth.

That GL(h) is a diffeomorphism boils down to the fact that the composite

$$\text{GL}(f)\text{GL}(h)\text{GL}(f)^{-1} \colon \text{GL}(\mathbf{R}^n) \to \text{GL}(\mathbf{R}^n)$$

is nothing but $\text{GL}(fhf^{-1})$. If $fhf^{-1} \colon \mathbf{R}^n \to \mathbf{R}^n$ is represented by the matrix A, then $\text{GL}(fhf^{-1})$ is represented by conjugation by A and hence a diffeomorphism.

If $\alpha, \beta \colon V \cong V$ are two linear isomorphisms, we may compose them to get $\alpha\beta \colon V \to V$. That GL($h$) respects composition follows, since $\text{GL}(h)(\alpha\beta) = h(\alpha\beta)h^{-1} = h\alpha h^{-1} h\beta)h^{-1} = \text{GL}(h)(\alpha)\text{GL}(h)(\beta)$. Also, GL($h$) preserves the identity element since $\text{GL}(h)(\text{id}_V) = h\,\text{id}_V h^{-1} = hh^{-1} = \text{id}_W$.

Exercise 2.5.12

Consider the chart $x \colon M_2(\mathbf{R}) \to \mathbf{R}^4$ given by

$$x\left(\begin{bmatrix} a & b \\ c & d \end{bmatrix}\right) = (a, b, a - d, b + c).$$

Exercise 2.5.13

The subset $f(\mathbf{RP}^n) = \{[p, 0] \in \mathbf{RP}^{n+1}\}$ is a submanifold by using all but the last of the standard charts on $\mathbf{RP}^{n+1}$. Checking that $\mathbf{RP}^n \to f(\mathbf{RP}^n)$ is a diffeomorphism is now straight-forward (the "ups, overs and acrosses" correspond to the chart transformations in $\mathbf{RP}^n$).

Exercise 2.5.16

Assume $i_j \colon N_j \to M_j$ are inclusions of submanifolds – the diffeomorphism part of "imbedding" being the trivial case – and let $x_j \colon U_j \to U_j'$ be charts such that

$$x_j(U_j \cap N_j) = U_j' \cap (\mathbf{R}^{n_j} \times \{0\}) \subseteq \mathbf{R}^{m_j}$$

for $j = 1, 2$. To check whether f is smooth at $p \in N_1$ it suffices to assert that $x_2 f x_1^{-1}|_{x_1(V)} = x_2 g x_1^{-1}|_{x_1(V)}$ is smooth at p where $V = U_1 \cap N_1 \cap g^{-1}(U_2)$, which is done by checking the higher-order partial derivatives in the relevant coordinates.

Exercise 2.5.17

Let $f \colon X \to Y$ and $g \colon Y \to Z$ be imbeddings. Then the induced map $X \to gf(X)$ is a diffeomorphism. Hence it suffices to show that the composite of inclusions of submanifolds is an inclusion of a submanifold. Let $X \subseteq Y \subset Z$ be inclusions of submanifolds (of dimension n, $n + k$ and $n + k + l$). Given $p \in X$, let $z \colon U \to U'$ be a chart on Z such that $z(U \cap Y) = (\mathbf{R}^{n+k} \times \{0\}) \cap U'$ and let $y \colon V \to V'$ be a chart on Y such that $y(V \cap X) = (\mathbf{R}^n \times \{0\}) \cap V'$ with $p \in U \cap V$. We may even assume (by shrinking the domains) that $V = Y \cap U$. Then

$$\left(\left(yz^{-1}|_{z(V)} \times \text{id}_{\mathbf{R}^l}\right)|_{U'} \circ z, U\right)$$

is a chart displaying X as a submanifold of Z.

Exercise 2.5.18 With the standard notation for the atlas on $\mathbf{RP}^n$, prove that $U^{kj} = H(m, n) \cap (U^k \times U^j)$ gives an open cover of $H(m, n)$ when $k < j$ varies. Define charts $x^{kj} : U^k \times U^j \to \mathbf{R}^{m+n}$ via $x^{kj}([p], [q]) = \left(x^k([p], x_k^j([q]), (1/p_k q_j) \sum_{i=0}^m p_i q_i\right)$, where $x_k^j([q]) = (1/q_j)(q_0, \ldots, \hat{q}_k, \ldots, \hat{q}_j, \ldots, q_n)$. Show that this is well defined and gives a smooth atlas for $\bigcup_{k<j}(U^k \times U^j)$ displaying $H(m, n)$ as a smooth submanifold.

Exercise 2.6.2 Check that all chart transformations are smooth.

Exercise 2.6.5 Up, over and across, using appropriate charts on the product, reduces this to saying that the projection $\mathbf{R}^m \times \mathbf{R}^n \to \mathbf{R}^m$ is smooth and that the inclusion of $\mathbf{R}^m$ in $\mathbf{R}^m \times \mathbf{R}^n$ is an imbedding.

Exercise 2.6.6 The heart of the matter is that $\mathbf{R}^k \to \mathbf{R}^m \times \mathbf{R}^n$ is smooth if and only if both the composites $\mathbf{R}^k \to \mathbf{R}^m$ and $\mathbf{R}^k \to \mathbf{R}^n$ are smooth.

Exercise 2.6.7 Consider the map $(t, z) \mapsto e^t z$.

Exercise 2.6.8 Reduce to the case where f and g are inclusions of submanifolds. Then rearrange some coordinates to show that case.

Exercise 2.6.9 Use the preceding exercises.

Exercise 2.6.10 Remember that $\mathrm{GL}_n(\mathbf{R})$ is an open subset of $M_n(\mathbf{R})$ and so this is in flatland, Example 2.5.8. Multiplication of matrices is smooth since it is made out of addition and multiplication of real numbers.

Exercise 2.6.11 Use the fact that multiplication of complex numbers is smooth, plus Exercise 2.5.16.

Exercise 2.6.13 Check chart transformations.

Exercise 2.6.17 Using the "same" charts on both sides, this reduces to saying that the identity is smooth.

Exercise 2.6.18 A map from a disjoint union is smooth if and only if it is smooth on both summands since smoothness is measured locally.

Exercise 2.6.19 It is a finite list. When you're done, please send it to me and proceed to the other chapters.

Chapter 3

Exercise 3.1.5

The only thing that is slightly ticklish with the definition of germs is the transitivity of the equivalence relation: assume

$$f : U_f \to N, \qquad g : U_g \to N, \text{ and } h : U_h \to N$$

and $f \sim g$ and $g \sim h$. On writing out the definitions, we see that $f = g = h$ on the open set $V_{fg} \cap V_{gh}$, which contains p.

Exercise 3.1.6

Choosing other representatives changes nothing in the intersection of the domains of definition. Associativity and the behavior of identities follow from the corresponding properties for the composition of representatives.

Exercise 3.1.8

If $\phi : U \to V$ is a diffeomorphism such that $f(t) = \phi(t)$ for all $t \in U$, then ϕ^{-1} represents an inverse to $\bar{f}$. Conversely, let $g : V_g \to M$ represent an inverse to $\bar{f}$. Then there is an open neighborhood $p \in U_{gf}$ such that $u = gf(u)$ for all $u \in U_{gf} \subseteq U_f \cap f^{-1}(V_g)$ and an open neighborhood $q \in V_{fg} \subseteq g^{-1}(U_f) \cap V_g$ such that $v = fg(v)$ for all $v \in V_{fg}$. Letting $U = U_{gf} \cap f^{-1}(V_{gf})$ and $V = g^{-1}(U_{gf}) \cap V_{gf}$, the restriction of f to U defines the desired diffeomorphism $\phi : U \to V$.

Exercise 3.1.12

Both sides send the function germ $\bar{\psi} : (L, g(f(p))) \to (\mathbf{R}, \psi(g(f(p))))$ to the composite

$$(M, p) \xrightarrow{\bar{f}} (N, f(p)) \xrightarrow{\bar{g}} (L, g(f(p)))$$
$$\downarrow{\bar{\psi}}$$
$$(\mathbf{R}, \psi(g(f(p)))),$$

i.e., $f^* g^*(\bar{\psi}) = f^*(\overline{\psi g}) = (\overline{\psi g})\bar{f} = \bar{\psi}(\overline{gf}) = (gf)^*(\bar{\psi})$.

Exercise 3.2.5

We do it for $\epsilon = \pi/2$. Other ϵs are then obtained by scaling. Let

$$f(t) = \gamma_{(\pi/4, \pi/4)}(t) \cdot t + (1 - \gamma_{(\pi/4, \pi/4)}(t)) \cdot \tan(t).$$

As to the last part, if $\bar{\gamma} : (\mathbf{R}, 0) \to (M, p)$ is represented by $\gamma_1 : (-\epsilon, \epsilon) \to M$, we let $\gamma = \gamma_1 f^{-1}$, where f is a diffeomorphism $(-\epsilon, \epsilon) \to \mathbf{R}$ with $f(t) = t$ for $|t|$ small.

Exercise 3.2.6

Let $\phi : U_\phi \to \mathbf{R}$ be a representative for $\bar{\phi}$, and let (x, U) be any chart around p such that $x(p) = 0$. Choose an $\epsilon > 0$ such that $x(U \cap U_\phi)$ contains the open ball of radius ϵ. Then the germ represented by ϕ is equal to the germ represented by the map defined on all of M given by

$$q \mapsto \begin{cases} \gamma_{(\epsilon/3, \epsilon/3)}(x(q))\phi(q) & \text{for } q \in U \cap U_\phi \\ 0 & \text{otherwise.} \end{cases}$$

Exercise 3.2.7

You can extend any chart to a function defined on the entire manifold.

Exercise 3.3.2

Let $\gamma, \gamma_1 \in W_p$. If (x, U) is a chart with $p \in U$ and if for all function germs $\bar{\phi} \in \mathcal{O}_p$ $(\phi\gamma)'(0) = (\phi\gamma_1)'(0)$, then by letting $\phi = x_k$ be the kth coordinate of x for $k = 1, \ldots, n$ we get that $(x\gamma)'(0) = (x\gamma_1)'(0)$. Conversely, assume $\bar{\phi} \in \mathcal{O}_p$ and $(x\gamma)'(0) = (x\gamma_1)'(0)$ for all charts (x, U). Then $(\phi\gamma)'(0) = (\phi x^{-1} x\gamma)'(0) = D(\phi x^{-1})(x(p)) \cdot (x\gamma)'(0)$ by the flat chain rule, Lemma 3.0.3, and we are done.

Exercise 3.3.3

If (y, V) is some other chart with $p \in V$, then the flat chain rule, Lemma 3.0.3, gives that

$$
\begin{aligned}
(y\gamma)'(0) &= (yx^{-1}x\gamma)'(0) \\
&= D(yx^{-1})(x(p)) \cdot (x\gamma)'(0) \\
&= D(yx^{-1})(x(p)) \cdot (x\gamma_1)'(0) \\
&= (yx^{-1}x\gamma_1)'(0) = (y\gamma_1)'(0),
\end{aligned}
$$

where $D(yx^{-1})(x(p))$ is the Jacobi matrix of the function yx^{-1} at the point $x(p)$.

Exercise 3.3.5

It depends neither on the representation of the tangent vector nor on the representation of the germ, because, if $[\gamma] = [v]$ and $\bar{f} = \bar{g}$, then $(\phi f\gamma)'(0) = (\phi f v)'(0) = (\phi g v)'(0)$ (partially by definition).

Exercise 3.3.8

This is immediately evident from the chain rule (or, for that matter, from the definition).

Exercise 3.3.13

$\alpha^{-1}(a\alpha(x) + b\alpha(y)) = \alpha^{-1}\beta^{-1}\beta(a\alpha(x) + b\alpha(y)) = (\beta\alpha)^{-1}(a\beta\alpha(x) + b\beta\alpha(y)).$

Exercise 3.3.16

Upon expanding along the ith row, we see that the partial differential of det with respect to the i, j-entry is equal to the determinant of the matrix you get by deleting the ith row and the jth column. Hence, the Jacobian of det is the $1 \times n^2$ matrix consisting of these determinants (in some order), and is zero if and only if all of them vanish, which is the same as saying that A has rank less than $n - 1$.

Exercise 3.3.17

This can be done either by sitting down and calculating partial derivatives or by arguing abstractly. Since the Jacobian $DL(p)$ represents the unique linear map K such that $\lim_{h\to 0}(1/h)(L(p + h) - L(p) - K(h)) = 0$ and L is linear, we get that $K = L$.

Exercise 3.4.3

This follows directly from the definition.

Exercise 3.4.6

Both ways around the square send $\bar{\phi} \in \mathcal{O}_{M,p}$ to $d(\phi f)$.

Exercise 3.4.13

If V has basis $\{v_1, \ldots, v_n\}$ and W has basis $\{w_1, \ldots, w_m\}$, then $f(v_i) = \sum_{j=1}^{m} a_{ij} w_j$ means that $A = (a_{ij})$ represents f in the given basis. Then $f^*(w_j^*) = w_j^* f = \sum_{i=1}^{n} a_{ij} v_i^*$, as can be checked by evaluating at v_i: $w_j^* f(v_i) = w_j^*(\sum_{k=1}^{m} a_{ik} w_k) = a_{ij}$.

Exercise 3.4.16

Use the definitions.

Exercise 3.4.19

The two Jacobi matrices in question are given by

$$D(xy^{-1})(y(p))^{\mathrm{T}} = \begin{bmatrix} p_1/|p| & p_2/|p| \\ -p_2 & p_1 \end{bmatrix}$$

and

$$D(yx^{-1})(x(p)) = \begin{bmatrix} p_1/|p| & -p_2/|p|^2 \\ p_2/|p| & p_1/|p|^2 \end{bmatrix}.$$

Exercise 3.5.6

Let $X : \mathcal{O}_{M,p} \to \mathbf{R}$ be a derivation, then

$$\begin{aligned} D|_{f(p)}g(D|_p f(X)) &= D|_{f(p)}g(Xf^*) \\ &= (Xf^*)g^* = X(gf)^* \\ &= D|_p gf(X). \end{aligned}$$

Exercise 3.5.11

Assume

$$X = \sum_{j=1}^{n} v_j \, D_j\big|_0 = 0.$$

Then

$$0 = X(\overline{\mathrm{pr}_i}) = \sum_{j=1}^{n} v_j D_j(\mathrm{pr}_i)(0) = \begin{cases} 0 & \text{if } i \neq j \\ v_i & \text{if } i = j. \end{cases}$$

Hence $v_i = 0$ for all i and we have linear independence.

If $X \in D|_0 \mathbf{R}^n$ is any derivation, let $v_i = X(\overline{\mathrm{pr}_i})$. If $\bar{\phi}$ is any function germ, we have by Lemma 3.4.8 that

$$\bar{\phi} = \phi(0) + \sum_{i=1}^{n} \overline{\mathrm{pr}_i} \cdot \overline{\phi_i}, \qquad \phi_i(p) = \int_0^1 D_i\phi(t \cdot p)dt,$$

and so

$$\begin{aligned} X(\bar{\phi}) &= X(\phi(0)) + \sum_{i=1}^{n} X(\overline{\mathrm{pr}_i} \cdot \overline{\phi_i}) \\ &= 0 + \sum_{i=1}^{n} \left(X(\overline{\mathrm{pr}_i}) \cdot \phi_i(0) + \mathrm{pr}_i(0) \cdot X(\overline{\phi_i}) \right) \\ &= \sum_{i=1}^{n} \left(v_i \cdot \phi_i(0) + 0 \cdot X(\overline{\phi_i}) \right) \\ &= \sum_{i=1}^{n} v_i D_i\phi(0), \end{aligned}$$

where the identity $\phi_i(0) = D_i\phi(0)$ was used in the last equality.

Exercise 3.5.15   If $[\gamma] = [\nu]$, then $(\phi\gamma)'(0) = (\phi\nu)'(0)$.

Exercise 3.5.18   The tangent vector $[\gamma]$ is sent to $X_\gamma f^*$ one way, and $X_{f\gamma}$ the other, and if we apply this to a function germ $\bar{\phi}$ we get

$$X_\gamma f^*(\bar{\phi}) = X_\gamma(\bar{\phi}\bar{f}) = (\phi f\gamma)'(0) = X_{f\gamma}(\bar{\phi}).$$

If you find such arguments hard, notice that $\phi f\gamma$ is the **only possible composition of these functions**, and so either side had better relate to this!

Chapter 4

Exercise 4.1.5   Using the identity charts, we get that the Jacobian is given by

$$Df(s, t) = \begin{bmatrix} 2s & 0 \\ t & s \end{bmatrix},$$

implying that

$$\mathrm{rk}_{(s,t)} f = \begin{cases} 0 & \text{if } s = t = 0 \\ 1 & \text{if } s = 0 \neq t \\ 2 & \text{if } s \neq 0. \end{cases}$$

Exercise 4.1.12   If $f : M \to N$ and $p \in M$, the diagram

$$
\begin{array}{ccc}
T^*_{f(p)}N & \xrightarrow{T^*_p f} & T^*_p M \\
{\scriptstyle \cong} \downarrow {\scriptstyle \alpha_{N,f(p)}} & & {\scriptstyle \cong} \downarrow {\scriptstyle \alpha_{M,p}} \\
(T_{f(p)}N)^* & \xrightarrow{(T_p f)*} & (T_p M)^*
\end{array}
$$

of Proposition 3.4.14 commutes. Now, the rank of the linear map $T_p f$ and its dual $(T_p f)^*$ agree (the rank is not affected by transposition of matrices).

Exercise 4.1.13   Observe that the function in question is

$$f(e^{i\theta}, e^{i\phi}) = \sqrt{(3 - \cos\theta - \cos\phi)^2 + (\sin\theta + \sin\phi)^2},$$

giving the claimed Jacobi matrix. Then solve the system of equations

$$3 \sin\theta - \cos\phi \sin\theta + \sin\phi \cos\theta = 0,$$
$$3 \sin\phi - \cos\theta \sin\phi + \sin\theta \cos\phi = 0.$$

By adding the two equations we get that $\sin\theta = \sin\phi$, but then the upper equation claims that $\sin\phi = 0$ or $3 - \cos\phi + \cos\theta = 0$. The latter is clearly impossible.

Exercise 4.2.5

Consider the smooth map

$$f: G \times G \to G \times G,$$
$$(g, h) \mapsto (gh, h)$$

with inverse $(g, h) \mapsto (gh^{-1}, h)$. Use that, for a given $h \in G$, the map $L_h: G \to G$ sending g to $L_h(g) = gh$ is a diffeomorphism, and that

$$
\begin{array}{ccc}
T_g G \times T_h G & \xrightarrow{\left[\begin{smallmatrix} T_g L_h & T_h R_g \\ 0 & 1 \end{smallmatrix}\right]} & T_{gh} G \times T_h G \\
\cong \Big\downarrow & & \cong \Big\downarrow \\
T_{(g,h)}(G \times G) & \xrightarrow{T_{(g,h)} f} & T_{(gh,h)}(G \times G)
\end{array}
$$

commutes (where the vertical isomorphisms are the "obvious" ones and $R_g(h) = gh$), to conclude that f has maximal rank, and is a diffeomorphism. Then consider a composite

$$G \xrightarrow{g \mapsto (1, g)} G \times G \xrightarrow{f^{-1}} G \times G \xrightarrow{(g,h) \mapsto g} G.$$

Perhaps a word about the commutativity of the above square is desirable. Starting with a pair $([\gamma], [\eta])$ in the upper left-hand corner, going down, right and up you get $([\gamma \cdot \eta], [\eta])$. However, if you go along the upper map you get $([\gamma \cdot h] + [g \cdot \eta], [\eta])$, so we need to prove that $[\gamma \cdot \eta] = [\gamma \cdot h] + [g \cdot \eta]$.

Choose a chart (z, U) around $g \cdot h$, let $\Delta: \mathbf{R} \to \mathbf{R} \times \mathbf{R}$ be given by $\Delta(t) = (t, t)$ and let $\mu: G \times G \to G$ be the multiplication in G. Then the chain rule, as applied to the function $z(\gamma \cdot \eta) = z\mu(\gamma, \eta)\Delta: \mathbf{R} \to \mathbf{R}^n$, gives that

$$(z(\gamma \cdot \eta))'(0) = D(z\mu(\gamma, \eta))(0, 0) \cdot \Delta'(0)$$

$$= \begin{bmatrix} D_1(z\mu(\gamma, \eta))(0, 0) & D_2(z\mu(\gamma, \eta))(0, 0) \end{bmatrix} \cdot \begin{bmatrix} 1 \\ 1 \end{bmatrix}$$

$$= (z(\gamma \cdot h))'(0) + (z(g \cdot \eta))'(0).$$

Exercise 4.3.2

Let $(t_1, \ldots, t_n) = e_1 = (1, 0, \ldots, 0)$. Then $t_{\sigma^{-1}(j)}$ is going to be 1 if $\sigma^{-1}(j) = 1$ (i.e., if $\sigma(1) = j$) and zero otherwise. Hence, the permutation of the coordinates sends e_1 to $(t_{\sigma^{-1}(1)}, \ldots, t_{\sigma^{-1}(n)}) = e_{\sigma(1)}$ – the $\sigma(1)$st standard unit vector. Likewise we get that the permutation of the coordinates sends the e_k to $e_{\sigma(k)}$, and the matrix associated with the permutation of the coordinates is $[e_{\sigma(1)}, \ldots, e_{\sigma(n)}]$.

Exercise 4.3.4

The rank theorem says that around any regular point there is a neighborhood on which f is a diffeomorphism. Hence $f^{-1}(q)$ is discrete, and, since M is compact, finite. Choose open neighborhoods U_x around each element in $x \in f^{-1}(q)$ such that f defines a diffeomorphism from U_x to an open neighborhood $f(U_x)$ of q. Let the promised neighborhood around q be

$$\bigcap_{x \in f^{-1}(q)} f(U_x) - f\left(M - \bigcup_{x \in f^{-1}(q)} U_x\right)$$

(remember that f takes closed sets to closed sets since M is compact and N Hausdorff).

Exercise 4.3.5

(After p. 8 of [15].) Extend P to a smooth map $f : S^2 \to S^2$ by stereographic projection (check that f is smooth at the North pole). Assume 0 is a regular value (if it is a critical value we are done!). The critical points of f correspond to the zeros of the derivative P', of which there are only finitely many. Hence the regular values of f cover all but finitely many points of S^2, and so give a connected space. Since by Exercise 4.3.4 $q \mapsto |f^{-1}(q)|$ is a locally constant function of the regular values, we get that there is an n such that $n = |f^{-1}(q)|$ for all regular values q. Since n can't be zero (P was not constant) we are done.

Exercise 4.3.6

(After [4].) Use the rank theorem, which gives the result immediately if we can prove that the rank of f is constant (spell this out). To prove that the rank of f is constant, we first prove it for all points in $f(M)$ and then extend it to some neighborhood using the chain rule.

The chain rule gives that

$$T_p f = T_p(ff) = T_{f(p)} f T_p f.$$

If $p \in f(M)$, then $f(p) = p$, so we get that $T_p f = T_p f T_p f$ and

$$T_p f(T_p M) = \{v \in T_p M \mid T_p f(v) = v\} = \ker\{1 - T_p f\}.$$

By the dimension theorem in linear algebra we get that

$$rk\left(T_p f\right) + rk\left(1 - T_p f\right) = \dim(M),$$

and since both ranks can only increase locally, they must be locally constant, and hence constant, say $rk T_p f = r$ and $rk(1 - T_p f) = \dim(M) - r$, since M was supposed to be connected.

Hence there is an open neighborhood U of $p \in f(M)$ such that $rk\, T_q f \geq r$ for all $q \in U$, but since $T_q f = T_{f(q)} f T_q f$ we must have $rk\, T_q f \leq T_{f(p)} f = r$, and so $rk\, T_q f = r$ too.

That $f(M) = \{p \in M \mid f(p) = p\}$ is closed in M follows since the complement is open: if $p \neq f(p)$ choose disjoint open sets U and V around p and $f(p)$. Then $U \cap f^{-1}(V)$ is an open set disjoint from $f(M)$ (since $U \cap f^{-1}(V) \subseteq U$ and $f(U \cap f^{-1}(V)) \subseteq V$) containing p.

Exercise 4.4.4

Prove that 1 is a regular value for the function $\mathbf{R}^{n+1} \to \mathbf{R}$ sending p to $|p|^2$.

Exercise 4.4.7

Show that the map

$$\text{SL}_2(\mathbf{R}) \to (\mathbf{C} \setminus \{0\}) \times \mathbf{R}$$

$$\begin{bmatrix} a & b \\ c & d \end{bmatrix} \mapsto (a + ic, ab + cd)$$

is a diffeomorphism, with inverse

$$(a + ic, t) \mapsto \begin{bmatrix} a & \frac{at-c}{a^2+c^2} \\ c & \frac{a+ct}{a^2+c^2} \end{bmatrix},$$

and that $S^1 \times \mathbf{R}$ is diffeomorphic to $\mathbf{C} \setminus \{0\}$.

Exercise 4.4.8

Calculate the Jacobi matrix of the determinant function. With some choice of indices you should get

$$D_{ij}(\det)(A) = (-1)^{i+j} \det(A_{ij}),$$

where A_{ij} is the matrix you get by deleting the ith row and the jth column. If the determinant is to be one, some of the entries in the Jacobi matrix then must be nonzero.

Exercise 4.4.11

Copy one of the proofs for the orthogonal group, replacing the symmetric matrices with Hermitian matrices.

Exercise 4.4.12

The space of orthogonal matrices is compact since it is a closed subset of $[-1, 1]^{n^2}$. It has at least two components since the set of matrices with determinant 1 is closed, as is the complement: the set with determinant -1.

Each of these is connected since you can get from any rotation to the identity through a path of rotations. One way to see this is to use the fact from linear algebra which says that any element $A \in SO(n)$ can be written in the form $A = BTB^{-1}$, where B and T are orthogonal, and furthermore T is a block diagonal matrix where the block matrices are either a single 1 on the diagonal, or of the form

$$T(\theta_k) = \begin{bmatrix} \cos\theta_k & -\sin\theta_k \\ \sin\theta_k & \cos\theta_k \end{bmatrix}.$$

So we see that by replacing all the θ_ks by $s\theta_k$ and letting s vary from 1 to 0 we get a path from A to the identity matrix.

Exercise 4.4.13

Elements of $SO(2)$ are of the form

$$\begin{bmatrix} a & b \\ c & d \end{bmatrix},$$

where $a^2 + c^2 = 1$, $ab + cd = 0$, $b^2 + d^2 = 1$ and $ad - bc = 1$. Conclude that $S^1 \to SO(2)$ sending e^{it} to

$$\begin{bmatrix} \cos t & -\sin t \\ \sin t & \cos t \end{bmatrix}$$

is a diffeomorphism.

Insofar as $SO(3)$ is concerned, identify $\mathbf{RP}^3$ with the quotient of the closed unit 3-ball where antipodal points on the boundary have been identified. Consider the function that sends a point p in this space to the element of $SO(3)$ corresponding to the rotation around the vector p by an angle $\pi|p|$. Beware of what happens when $|p| = 1$. Check that this is a smooth bijection inducing an isomorphism on each tangent space. It is probably useful to write out a formula.

Exercise 4.4.16

Consider a k-frame as a matrix A with the property that $A^TA = I$, and proceed as for the orthogonal group.

Exercise 4.4.18

Either just solve the equation or consider the map

$$f : P_3 \to P_2$$

sending $y \in P_3$ to $f(y) = (y'')^2 - y' + y(0) + xy'(0) \in P_2$. If you calculate the Jacobian in obvious coordinates you get that

$$Df(a_0 + a_1 x + a_2 x^2 + a_3 x^3) = \begin{bmatrix} 1 & -1 & 8a_2 & 0 \\ 0 & 1 & 24a_3 - 2 & 24a_2 \\ 0 & 0 & 0 & 72a_3 - 3 \end{bmatrix}.$$

The only way this matrix can be singular is if $a_3 = 1/24$, but the top coefficient in $f(a_0 + a_1 x + a_2 x^2 + a_3 x^3)$ is $36a_3^2 - 3a_3$, which won't be zero if $a_3 = 1/24$. By the way, if I did not calculate something wrong, the solution is the disjoint union of two manifolds $M_1 = \{2t(1 - 2t) + 2tx + tx^2 \,|\, t \in \mathbf{R}\}$ and $M_2 = \{-24t^2 + tx^2 + x^3/12 \,|\, t \in \mathbf{R}\}$, both of which are diffeomorphic to $\mathbf{R}$.

Exercise 4.4.19

Yeah.

Exercise 4.4.20

Consider the function

$$f : \mathbf{R}^n \to \mathbf{R}$$

given by

$$f(p) = p^{\mathrm{T}} A p.$$

The Jacobi matrix is easily calculated, and using that A is symmetric we get that $Df(p) = 2p^{\mathrm{T}} A$. Hence, given that $f(p) = b$, we get that $\frac{1}{2} Df(p) \cdot p = p^{\mathrm{T}} A p = b$, and so $Df(p) \neq 0$ if $b \neq 0$. Hence all values but $b = 0$ are regular. The value $b = 0$ is critical since $0 \in f^{-1}(0)$ and $Df(0) = 0$.

Exercise 4.4.21

You don't actually need Theorem 4.4.3 to prove this since you can isolate T in this equation, and show directly that you get a submanifold diffeomorphic to $\mathbf{R}^2$, but still, as an exercise you should do it by using Theorem 4.4.3.

Exercise 4.4.22

Code a flexible n-gon by means of a vector $x^0 \in \mathbf{R}^2$ giving the coordinates of the first point, and vectors $x^i \in S^1$ going from point i to point $i + 1$ for $i = 1, \ldots, n - 1$ (the vector from point n to point 1 is not needed, since it will be given by the requirement that the curve is closed). The set $\mathbf{R}^2 \times (S^1)^{n-1}$ will give a flexible n-gon, except that the last line may not be of length 1. To ensure this, look at the map

$$f : \mathbf{R}^2 \times (S^1)^{n-1} \to \mathbf{R},$$

$$(x^0, (x^1, \ldots, x^{n-1})) \mapsto \left| \sum_{i=1}^{n-1} x^i \right|^2$$

and show that 1 is a regular value. If you let $x^j = e^{i\phi_j}$ and $x = (x^0, (x^1, \ldots, x^{n-1}))$, you get that

$$D_j f(x) = D_j \left(\left(\sum_{k=1}^{n-1} e^{i\phi_k} \right) \left(\sum_{k=1}^{n-1} e^{-i\phi_k} \right) \right)$$

$$= i e^{i\phi_j} \left(\sum_{k=1}^{n-1} e^{-i\phi_k} \right) + \left(\sum_{k=1}^{n-1} e^{i\phi_k} \right) (-i e^{-i\phi_j}).$$

That the rank is not 1 is equivalent to $D_j f(x) = 0$ for all j. Analyzing this, we get that $x_1, \ldots, x_{n-1}$ must then all be parallel. But this is impossible if n is odd and $\left| \sum_{i=1}^{n-1} x^i \right|^2 = 1$. (Note that this argument fails for n even. If $n = 4$ then $F_{4,2}$ is not a manifold: given x^1 and x^2 there are two choices for x^3 and x^4 (either $x^3 = -x^2$ and $x^4 = -x^1$ or $x^3 = -x^1$ and $x^4 = -x^2$), but when $x^1 = x^2$ we get a crossing of these two choices.)

Exercise 4.4.23

The non-self-intersecting flexible n-gons form an open subset.

Exercise 4.5.3

Under the identification $T_z \mathbf{R}^2 \cong \mathbf{R}^2$ given by $[\gamma] \mapsto \gamma'(0)$, $T_z S^1$ corresponds to $\{v \mid z \cdot v = 0\}$ and $T_z N_z$ to $\{(0, y) \mid y \in \mathbf{R}\}$. These subspaces span $\mathbf{R}^2$ except when $z = (\pm 1, 0)$.

Exercise 4.5.7

Before we start, it is perhaps smart to review the relation between derivations in the real case and those in the complex case. If $g = u + iv$ represents a function $\mathbf{C} \to \mathbf{C}$ in standard complex analysis notation (which we will use freely for a short while), the usual identification $\mathbf{C} \cong \mathbf{R}^2$ displays the Jacobian as

$$Dg(z) = \begin{bmatrix} \partial u / \partial x & \partial u \partial y \\ \partial v / \partial x & \partial v \partial y \end{bmatrix}.$$

Under the identification in Exercise 2.5.12 of a complex number $z = x + iy$ with real matrix

$$\begin{bmatrix} x & -y \\ y & x \end{bmatrix},$$

we see that g satisfies the Cauchy–Riemann equations *exactly* when $Dg(z)$ corresponds to a complex number, which in this case is nothing but the complex derivative $g'(z)$. In particular, if $g(z) = z^a$ for some positive integer a, then $Dg(z)$ corresponds to $g'(z) = az^{a-1}$, and so has rank 0 if $z = 0$ and rank 2 otherwise.

Getting back to the exercise, we first prove that $L = f^{-1}(0) \neq \emptyset$ is a submanifold by proving that 0 is a regular value for f: on representing $x + iy \in \mathbf{C}$ by the real matrix

$$\begin{bmatrix} x & -y \\ y & x \end{bmatrix},$$

we can express the Jacobian as

$$Df(z) = [a_0 z_0^{a_0 - 1}, \ldots, a_n z_n^{a_n - 1}],$$

which has maximal rank 2 since $(z_0, \ldots, z_n) \neq 0$.

To see that L and S^{2n+1} are transverse in $z \in L \cap S^{2n+1}$, it suffices to demonstrate that L has a tangent vector $[\gamma]$ at z which is not in $T_z S^{2n+1}$, i.e., such that $\gamma'(0) \cdot z \neq 0$. Letting a be the least common multiple of $a_0, \ldots, a_n$, the curve given by

$$\gamma(t) = ((1+t)^{a/a_0} z_0, \ldots, (1+t)^{a/a_n} z_n) \in L$$

has

$$\gamma'(0) = \left(\frac{a}{a_0} z_0, \ldots, \frac{a}{a_n} z_n \right),$$

and so

$$\gamma'(0) \cdot z = \sum_k \frac{a}{a_k} |z_k|^2 > 0.$$

Exercise 4.6.2

It suffices to prove that for any point $p \in U$ there is a point $q \in U$ with all coordinates rational and a rational r such that $p \in C \subseteq U$ with C the closed ball with center q and radius r. Since U is open, there is an $\epsilon > 0$ such that the open ball with center p and radius ϵ is within U. Since $\mathbf{Q}^n \subseteq \mathbf{R}^n$ is dense we may choose $r \in \mathbf{Q}$ and $q \in \mathbf{Q}^n$ such that $|q - p| < r < \epsilon/2$.

Exercise 4.6.3

Let $\{C_i\}_{i \in \mathbf{N}}$ be a countable collection of measure-zero sets, let $\epsilon < 0$ and for each $i \in \mathbf{N}$ choose a sequence of cubes $\{C_{ij}\}_{j \in \mathbf{N}}$ with $C_i \subseteq \bigcup_{j \in \mathbf{N}} C_{ij}$ and $\sum_{j \in \mathbf{N}} \text{volume}(C_{ij}) < \epsilon/2^i$.

Exercise 4.6.4

By Exercise 4.6.2 we may assume that C is contained in a closed ball contained in U. On choosing $\epsilon > 0$ small enough, a covering of C by closed balls $\{C_i\}$ whose sum of volumes is less than ϵ will also be contained in a closed ball K contained in U. Now, the mean value theorem assures that there is a positive number M such that $|f(a) - f(b)| \leq M|a - b|$ for $a, b \in K$. Hence, f sends closed balls of radius r into closed balls of radius Mr, and $f(C)$ is covered by closed balls whose sum of volumes is less than $M\epsilon$.

Note the crucial importance of the mean value theorem. The corresponding statements are false if we just assume our maps are continuous.

Exercise 4.6.6

Since $[0, 1]$ is compact, we may choose a finite subcover. By excluding all subintervals contained in another subinterval we can assure that no point in $[0, 1]$ lies in more than two subintervals (the open cover $\{[0, 1), (0, 1]\}$ shows that 2 is attainable).

Exercise 4.6.7

It suffices to do the case where C is compact, and we may assume that $C \subseteq [0, 1]^n$. Let $\epsilon > 0$. Given $t \in [0, 1]$, let $d_t : C \to \mathbf{R}$ be given by $d_t(t_1, \ldots, t_n) = |t_n - t|$ and let $C^t = d_t^{-1}(0)$. Choose a cover $\{B_i^t\}$ of C^t by open cubes whose sum of volumes is less than $\epsilon/2$. Let $J_t : \mathbf{R}^{n-1} \to \mathbf{R}^n$ be given by $J_t(t_1, \ldots, t_{n-1}) = (t_1, \ldots, t_{n-1}, t)$ and $B^t = J_t^{-1}(\bigcup_i B_i^t)$. Since C is compact and B^t is open, d_t attains a minimum value $m_t > 0$ outside $B^t \times \mathbf{R}$, and so $d_t^{-1}(-m_t, m_t) \subseteq B^t \times I_t$, where $I_t = (t - m_t, t + m_t) \cap [0, 1]$. By Exercise 4.6.6, there is a finite collection $\{t_1, \ldots, t_k\} \in [0, 1]$ such that the I_{t_j} cover $[0, 1]$ and such that the sum of the diameters is less than 2. From this we get the cover of C by rectangles $\{B_i^{t_j} \times I_{t_j}\}_{j=1,\ldots,k, i \in \mathbf{N}}$ whose sum of volumes is less than $\epsilon = 2\epsilon/2$.

Exercise 4.6.8　Use the preceding string of exercises.

Exercise 4.6.9　Let $C' = C_0 - C_1$. We may assume that $C' \neq \emptyset$ (excluding the case $m \leq 1$). If $p \in C'$, there is a nonzero partial derivative of f at p, and by permuting the coordinates, we may equally well assume that $D_1 f(p) \neq 0$. By the inverse function theorem, the formula

$$x(q) = (f_1(q), q_2, \ldots, q_m)$$

defines a chart $x: V \to V'$ in a neighborhood V of p, and it suffices to prove that $g(K)$ has measure zero, where $g = fx^{-1}: V' \to \mathbf{R}^n$ and K is the set of critical points for g. Now, $g(q) = (q_1, g_2(q), \ldots, g_n(q))$, and on writing $g_k^{q_1}(q_2, \ldots, q_m) = g_k(q_1, \ldots, q_m)$ for $k = 1, \ldots, n$, we see that since

$$Dg(q_1, \ldots, q_m) = \begin{bmatrix} 1 & 0 \\ ? & Dg^{q_1}(q_2, \ldots, q_m) \end{bmatrix}$$

the point $(q_1, \ldots, q_m)$ is a critical point for g if and only if $(q_2, \ldots, q_m)$ is a critical point for g^{q_1}. By the induction hypothesis, for each q_1 the set of critical values for g^{q_1} has measure zero. By Fubini's theorem (Exercise 4.6.7), we are done.

Exercise 4.6.10　The proof is similar to that of Exercise 4.6.9, except that the chart x is defined by

$$x(q) = (D_{k_1} \ldots D_{k_i} f(q), q_2, \ldots, q_m),$$

where we have assumed that

$$D_1 D_{k_1} \ldots D_{k_i} f(p) \neq 0$$

(but, of course $D_{k_1} \ldots D_{k_i} f(p) = 0$).

Exercise 4.6.11　By Exercise 4.6.2 U is a countable union of closed cubes (balls or cubes have the same proof), so it suffices to show that $f(K \cap C_k)$ has measure zero, where K is a closed cube with side s. Since all partial derivatives of order less than or equal to k vanish on C_k, Taylor expansion gives that there is a number M such that

$$|f(a) - f(b)| \leq M|a - b|^{k+1}$$

for all $a \in K \cap C_k$ and $b \in K$. Subdivide K into N^m cubes $\{K_{ij}\}_{i,j=1,\ldots,N}$ with sides s/N for some positive integer. If $a \in C_k \cap K_{ij}$, then $f(K_{ij})$ lies in a closed ball centered at $f(a)$ with radius $M(\sqrt{m} \cdot s/N)^{k+1}$. Consequently, $f(K \cap C_k)$ lies in a union of closed balls with volume sum less than or equal to

$$N^m \cdot \frac{4\pi \left(M(\sqrt{m} \cdot s/N)^{k+1}\right)^n}{3} = \frac{4\pi \left(M(\sqrt{m} \cdot s)^{k+1}\right)^n}{3} N^{m-n(k+1)}.$$

If $nk \geq m$, this tends to zero as N tends to infinity, and we are done.

Exercise 4.6.12　This is done by induction on m. When $m = 0$, $\mathbf{R}^m$ is a point and the result follows. Assume Sard's theorem is proven in dimension less than $m > 0$. Then Exercises 4.6.8, 4.6.9, 4.6.10 and 4.6.11 together prove Sard's theorem in dimension m.

Exercise 4.7.2
It is clearly injective, and an immersion since it has rank 1 everywhere. It is not an imbedding since $\mathbf{R} \coprod \mathbf{R}$ is disconnected, whereas the image is connected.

Exercise 4.7.3
It is clearly injective, and an immersion since it has rank 1 everywhere. It is not an imbedding since an open set containing a point z with $|z| = 1$ in the image must contain elements in the image of the first summand.

Exercise 4.7.6
Let $f \colon M_0 \to M_1$ and $g \colon M_1 \to M_2$ be imbeddings. You need only verify that $M_0 \to gf(M_0)$ is a homeomorphism and that for all $p \in M_0$ the tangent map $T_p gf = T_{f(p)} g T_p f$ is an injection.

Exercise 4.7.7
If a/b is irrational then the image of $f_{a,b}$ is dense; that is, any open set on $S^1 \times S^1$ intersects the image of $f_{a,b}$.

Exercise 4.7.8
Show that it is an injective immersion homeomorphic to its image. The last property follows since both the maps in

$$M \longrightarrow i(M) \longrightarrow ji(M)$$

are continuous and bijective and the composite is a homeomorphism.

Exercise 4.7.10
Prove that the diagonal $M \to M \times M$ is an imbedding by proving that it is an immersion inducing a homeomorphism onto its image. The tangent space of the diagonal at (p, p) is exactly the diagonal of $T_{(p,p)}(M \times M) \cong T_p M \times T_p M$.

Exercise 4.7.11
Show that the map

$$f \times g \colon M \times L \to N \times N$$

is transverse to the diagonal (which is discussed in Exercise 4.7.10). Identifying tangent spaces of products with products of tangent spaces, we must show that any $(v_1, v_2) \in T_r N \times T_r N$, where $r = f(p) = g(q)$, is of the form $(T_p f w_1, T_q g w_2) + (w, w)$. This is achieved by choosing w_1 and w_2 so that $v_2 - v_1 = T_q g w_2 - T_p f w_1$ (which is possible by the transversality hypothesis) and letting $w = v_1 - T_p f w_1$. Finally, show that the inverse image of the diagonal is exactly $M \times_N L$.

Exercise 4.7.12
Since π is a submersion, Exercise 4.7.11 shows that $E \times_M N \to N$ is smooth and that $E \times_M N \subseteq E \times N$ is a smooth submanifold. If $(e, n) \in E \times_M N$, with $\pi(e) = f(n) = m$, notice that the injection $T_{(e,n)}(E \times_M N) \to T_{(e,n)}(E \times N) \cong T_e E \times T_n N$ factors as

$$T_{(e,n)}(E \times_M N) \to T_e E \times_{T_m M} T_n N \subseteq T_e E \times T_n N.$$

Hence the map $T_{(e,n)}(E \times_M N) \to T_e E \times_{T_m M} T_n N$ is an injection, and – since the dimensions agree – an isomorphism. Finally, the map $T_{(e,n)}(E \times_M N) \to T_n N$ factors as

$$T_{(e,n)}(E \times_M N) \cong T_e E \times_{T_m M} T_n N \to T_n N,$$

where the last map is the projection, which is surjective since $T_e \pi \colon T_e E \to T_m M$ is.

Chapter 5

Exercise 5.1.4

For the first case, you may assume that the regular value in question is 0. Since zero is a regular value, the derivative in the "fiber direction" must be nonzero, and so the values of f are positive on one side of the zero section ... but there IS no "one side" of the zero section! This takes care of all one-dimensional cases, and higher-dimensional examples are excluded since the map won't be regular if the dimension increases.

Exercise 5.2.4

See the next exercise. This refers the problem away, but the same information helps you out on this one too!

Exercise 5.2.5

This exercise is solved in the smooth case in Exercise 5.3.15. The only difference in the continuous case is that you delete every occurrence of "smooth" in the solution. In particular, the solution refers to a "smooth bump function $\phi\colon U_2 \to \mathbf{R}$ such that ϕ is one on (a, c) and zero on $U_2 \setminus (a, d)$". This can in our case be chosen to be the (non-smooth) map $\phi\colon U_2 \to \mathbf{R}$ given by

$$\phi(t) = \begin{cases} 1 & \text{if } t \leq c \\ (d - t)/(d - c) & \text{if } c \leq t \leq d \\ 0 & \text{if } t \geq d. \end{cases}$$

Exercise 5.3.10

By Lemma 5.2.2 the transition functions are smooth if and only if the bundle chart transformations are smooth, and the smooth structure on E provided by Lemma 5.3.8 is exactly such that the bundle chart transformations are smooth.

Exercise 5.3.13

Check locally by using charts: if (h, U) is a bundle chart, then the resulting square

$$\begin{array}{ccc} E|U & \xrightarrow[\cong]{h} & U \times \mathbf{R}^k \\ a_E \downarrow & & \downarrow \mathrm{id}_U \times a\cdot \\ E|U & \xrightarrow[\cong]{h} & U \times \mathbf{R}^k \end{array}$$

commutes.

Exercise 5.3.14

Modify your solution to Exercise 5.2.4 so that it uses only smooth functions, or use parts of the solution of Exercise 5.3.15.

Exercise 5.3.15

Let $\pi\colon E \to S^1$ be a one-dimensional smooth line bundle. Since S^1 is compact we may choose a finite bundle atlas, and we may remove superfluous bundle charts, so that no domain is included in another. We may also assume that all chart domains are connected. If there is just one bundle chart, we are finished; otherwise we proceed as follows. If we start at some point, we may order the charts, so that they intersect in a nonempty interval (or a disjoint union of two intervals if there are exactly two charts).

Consider two consecutive charts (h_1, U_1) and (h_2, U_2), and let (a, b) be (one of the components of) their intersection. The transition function

$$h_{12} \colon (a, b) \to \mathbf{R} \setminus \{0\} \cong \mathrm{GL}_1(\mathbf{R})$$

must take either just negative or just positive values. On multiplying h_2 by the sign of h_{12} we get a situation where we may assume that h_{12} always is positive. Let $a < c < d < b$, and choose a smooth bump function $\phi \colon U_2 \to \mathbf{R}$ such that ϕ is one on (a, c) and zero on $U_2 \setminus (a, d)$. Define a new chart (h'_2, U_2) by letting

$$h'_2(t) = \left(\frac{\phi(t)}{h_{12}(t)} + 1 - \phi(t) \right) h_2(t)$$

(since $h_{12}(t) > 0$, the factor by which we multiply $h_2(t)$ is never zero). On (a, c) the transition function is now constantly equal to one, so if there were more than two charts we could merge our two charts into a chart with chart domain $U_1 \cup U_2$.

So we may assume that there are just two charts. Then we may proceed as above on one of the components of the intersection between the two charts, and get the transition function to be the identity. But then we would not be left with the option of multiplying with the sign of the transition function on the other component. However, by the same method, we could make it plus **or** minus one, which exactly correspond to the trivial bundle and the unbounded Möbius band.

Exactly the same argument shows that there are exactly two isomorphism types of rank-n smooth vector bundles over S^1 (using that $\mathrm{GL}_n(\mathbf{R})$ has exactly two components). The same argument also gives the corresponding topological fact.

Exercise 5.4.4

Use the chart domains on $\mathbf{RP}^n$ from the manifold section,

$$U^k = \{ [p] \in \mathbf{RP}^n \mid p_k \neq 0 \},$$

and construct bundle charts $\pi^{-1}(U^k) \to U^k \times \mathbf{R}$ sending $([p], \lambda p)$ to $([p], \lambda p_k)$. The chart transformations then should look something like

$$([p], \lambda) \mapsto \left([p], \lambda \frac{p_l}{p_k} \right).$$

If the bundle were trivial, then $\eta_n \setminus \sigma_0(\mathbf{RP}^n)$ would be disconnected. In particular, $([e_1], e_1)$ and $([e_1], -e_1)$ would be in different components. But $\gamma \colon [0, \pi] \to \eta_n \setminus \sigma_0(\mathbf{RP}^n)$ given by

$$\gamma(t) = ([\cos(t)e_1 + \sin(t)e_2], \cos(t)e_1 + \sin(t)e_2)$$

is a path connecting them.

Exercise 5.4.5

You may assume that $p = [0, \ldots, 0, 1]$. Then any point $[x_0, \ldots, x_{n-1}, x_n] \in X$ equals $[x/|x|, x_n/|x|]$ since $x = (x_0, \ldots, x_{n-1})$ must be different from 0. Consider the map

$$X \to \eta_{n-1},$$

$$[x, x_n] \mapsto \left(\left[\frac{x}{|x|} \right], \frac{x_n x}{|x|^2} \right)$$

with inverse $([x], \lambda x) \mapsto [x, \lambda]$.

Exercise 5.4.6

Let $\pi: E \to X$ be the projection and $\mathcal{B}$ be the bundle atlas of the first pre-vector bundle, and correspondingly for the other. The first requirement is that $\tilde{f}$ must induce a linear map on each fiber. This implies that $f(\pi(v)) = \pi'(\tilde{f}(v))$ gives a well-defined function $f: X \to X'$, and we must require that it is continuous. Lastly, for any bundle charts $(h, U) \in \mathcal{B}$, $(h', U') \in \mathcal{B}'$ we must have that the function $U \times \mathbf{R}^k \to U' \times \mathbf{R}^{k'}$ sending (p, v) to $h'\tilde{f}h^{-1}(p, v)$ is continuous.

Exercise 5.5.9

View S^3 as the unit quaternions, and copy the argument for S^1.

Exercise 5.5.10

A Lie group is a smooth manifold equipped with a smooth associative multiplication, having a unit and possessing all inverses, so the proof for S^1 will work.

Exercise 5.5.14

If we set $z_j = x_j + iy_j$, $x = (x_0, \ldots, x_n)$ and $y = (y_0, \ldots, y_n)$, then $\sum_{i=0}^n z^2 = 1$ is equivalent to $x \cdot y = 0$ and $|x|^2 - |y|^2 = 1$. Use this to make an isomorphism to the bundle in Example 5.5.11 sending the point (x, y) to $(p, v) = (x/|x|, y)$ (with inverse sending (p, v) to $(x, y) = (\sqrt{1 + |v|^2}p, v)$).

Exercise 5.5.16

Explicitly, if $s_1, s_2 \in \mathbf{R}$ and $\sigma_1, \sigma_2 \in \mathcal{X}(M)$ one observes that $s_1\sigma_1 + s_2\sigma_2: M \to TM$ given by $(s_1\sigma_1 + s_2\sigma_2)(p) = s_1\sigma_1(p) + s_2\sigma_2(p) \in T_pM$ is smooth. More generally, if $f_1, f_2 \in C^\infty(M)$, then we define $f_1\sigma_1 + f_2\sigma_2: M \to TM$ by $(f_1\sigma_1 + f_2\sigma_2)(p) = f_1(p)\sigma_1(p) + f_2(p)\sigma_2(p) \in T_pM$.

Exercise 5.5.17

Using the global trivialization of $TM \to M$, we may identify $\mathcal{X}(M)$ with the space of sections of the product bundle $M \times \mathbf{R}^n \to M$. A section of the product bundle is uniquely given by a map $M \to \mathbf{R}^n$ to the fiber, which is the same as an n-tuple of functions $M \to \mathbf{R}$.

Exercise 5.5.18

Identify $TS^n \to S^n$ with the bundle

$$E = \{(p, v) \in S^n \times \mathbf{R}^{n+1} \mid p \cdot v = 0\} \to S^n,$$

consider the map

$$\{\sigma \in C^\infty(S^n, \mathbf{R}^{n+1}) \mid p \cdot \sigma(p) = 0\} \times C^\infty(S^n) \to C^\infty(S^n, \mathbf{R}^{n+1})$$

sending (σ, f) to $p \mapsto \sigma(p) + f(p)p$ and use Exercise 2.6.6.

Exercise 5.5.19

Consider the isomorphism

$$TS^n \cong \{(p, v) \in S^n \times \mathbf{R}^{n+1} \mid p \cdot v = 0\}.$$

Any path $\bar{\gamma}$ in $\mathbf{R}\mathrm{P}^n$ through $[p]$ lifts uniquely to a path γ through p and to the corresponding path $-\gamma$ through $-p$.

Exercise 5.5.20

By Corollary 5.5.13 we identify $TO(n)$ with

$$E = \left\{ (g, A) \in O(n) \times M_n(\mathbf{R}) \; \middle| \; \begin{array}{l} g = \gamma(0) \\ A = \gamma'(0) \\ \text{for some curve} \\ \gamma : (-\epsilon, \epsilon) \to O(n) \end{array} \right\}.$$

That $\gamma(s) \in O(n)$ is equivalent to saying that $I = \gamma(s)^{\mathrm{T}} \gamma(s)$. This holds for all $s \in (-\epsilon, \epsilon)$, so we may derive this equation and get

$$\begin{aligned} 0 &= \frac{d}{ds}\Big|_{s=0} \left(\gamma(s)^{\mathrm{T}} \gamma(s) \right) \\ &= \gamma'(0)^{\mathrm{T}} \gamma(0) + \gamma(0)^{\mathrm{T}} \gamma'(0) \\ &= A^{\mathrm{T}} g + g^{\mathrm{T}} A. \end{aligned}$$

Exercise 5.5.22

Use the trivialization to pass the obvious solution on the product bundle to the tangent bundle.

Exercise 5.5.23

Any curve to a product is given uniquely by its projections to the factors.

Exercise 5.5.25

Let $x : U \to U'$ be a chart for M. Show that the assignment sending an element $(q, v_1, v_2, v_{12}) \in U' \times \mathbf{R}^n \times \mathbf{R}^n \times \mathbf{R}^n$ to

$$[t \mapsto [s \mapsto x^{-1}(q + tv_1 + sv_2 + stv_{12})]] \in T(TU)$$

gives an isomorphism

$$U' \times \mathbf{R}^n \times \mathbf{R}^n \times \mathbf{R}^n \cong T(TU),$$

so that all elements in $T(TM)$ are represented by germs of surfaces. Check that the equivalence relation is the one given in the exercise so that the resulting isomorphisms $T(TU) \cong E|_U$ give smooth bundle charts for E.

Exercise 5.6.1

Check that each of the pieces of the definition of a pre-vector bundle is accounted for.

Exercise 5.6.6

Choose a chart (x, U) with $p \in U$ and write out the corresponding charts on T^*M and $T^*(T^*M)$ to check smoothness. It may be that you will find it easier to think in terms of the "dual bundle" $(TM)^*$ rather than the isomorphic cotangent bundle T^*M and avoid the multiple occurrences of the isomorphism α, but strictly speaking the dual bundle will not be introduced until Example 6.4.8.

Chapter 6

Exercise 6.1.4

As an example, consider the open subset $U^{0,0} = \{e^{i\theta} \in S^1 \mid \cos\theta > 0\}$. The bundle chart $h\colon U^{0,0} \times \mathbf{C} \to U^{0,0} \times \mathbf{C}$ is given by sending $(e^{i\theta}, z)$ to $(e^{i\theta}, e^{-i\theta/2}z)$. Then $h((U^{0,0} \times \mathbf{C}) \cap \eta_1) = U^{0,0} \times \mathbf{R}$. Continue in this way all around the circle.

The idea is the same for higher dimensions: locally you can pick the first coordinate to be on the line $[p]$. For future reference we give explicit charts. Let $\{e_0, \ldots, e_n\}$ be the standard basis for $\mathbf{R}^{n+1}$. Given $p \in \mathbf{R}^{n+1}$ with $p_k \neq 0$, consider the basis $\{\tilde{e}_j\}$ with $\tilde{e}_k = (p_k/|p_k p|)p$ and $\tilde{e}_j = e_j - (e_j \cdot \tilde{e}_k)\tilde{e}_k$ for $k \neq j$. If $v \in \mathbf{R}^{n+1}$, we see that $v \cdot \tilde{e}_k = p_k v \cdot p/|p_k p|$ and $v \cdot \tilde{e}_j = v_j - p_j v \cdot p/|p|^2$ for $k \neq j$. We write $h_{[p]}v = (v \cdot \tilde{e}_0, \ldots, v \cdot \tilde{e}_n)$ and notice that $h_{[p]}$ is a linear isomorphism depending smoothly on $[p] \in U^k$. Furthermore, $h_{[p]}$ induces an isomorphism between the line $[p]$ and the line $[e_k]$. This allows us to give $\mathbf{RP}^n \times \mathbf{R}^{n+1} \to \mathbf{RP}^n$, the bundle chart over $U^k \subseteq \mathbf{RP}^n$, sending $([p], v) \in U^k \times \mathbf{R}^{n+1}$ to $([p], h_{[p]}v) \in U^k \times \mathbf{R}^{n+1}$, displaying (after permuting coordinates so that the kth coordinate becomes the first) η_n as a smooth subbundle of the product bundle. Note that the smooth structure on η_n agrees with the standard one.

Exercise 6.1.11

Let $X_k = \{p \in X \mid \mathrm{rk}_p f = k\}$. We want to show that X_k is both open and closed, and hence either empty or all of X since X is connected.

Let $P = \{A \in M_m(\mathbf{R}) \mid A = A^2\}$, i.e., the space of all projections. If $A \in P$, then the column space is the eigenspace corresponding to the eigenvalue 1 and the orthogonal complement is the eigenspace of the eigenvalue 0. Consequently,

$$P_k = \{A \in P \mid \mathrm{rk}\, A = k\}$$
$$= P \cap \{A \mid \mathrm{rk}\, A \geq k\} \cap \{A \mid \mathrm{rk}(A - I) \geq n - k\}$$

is open in P. However, given a bundle chart (h, U), the map

$$U \xrightarrow{p \mapsto h_p f_p h_p^{-1}} P$$

is continuous, and hence $U \cap X_k$ is open in U. By varying (h, U) we get that X_k is open, and hence also closed since $X_k = X \setminus \bigcup_{i \neq k} X_i$.

Exercise 6.1.12

Use Exercise 6.1.11 to show that the bundle map $\frac{1}{2}(\mathrm{id}_E - f)$ has constant rank (here we use that the set of bundle morphisms is in an obvious way a vector space).

Exercise 6.1.13

Identifying $T\mathbf{R}$ with $\mathbf{R} \times \mathbf{R}$ in the usual way, we see that f corresponds to $(p, v) \mapsto (p, p \cdot v)$, which is a nice bundle morphism, but

$$\coprod_p \ker\{v \mapsto pv\} = \{(p, v) \mid p \cdot v = 0\}$$

and

$$\coprod_p \mathrm{Im}\{v \mapsto pv\} = \{(p, pv)\}$$

are not subbundles.

Exercise 6.1.14   The tangent map $Tf: TE \to TM$ is locally the projection $U \times \mathbf{R}^k \times \mathbf{R}^n \times \mathbf{R}^k \to U \times \mathbf{R}^n$ sending (p, u, v, w) to (p, v), and so has constant rank. Hence Corollary 6.1.10 gives that $V = \ker\{Tf\}$ is a subbundle of $TE \to E$.

Exercise 6.2.5
$$A \times_X E = \pi^{-1}(A).$$

Exercise 6.2.6   This is not as complex as it seems. For instance, the map $\tilde{E} \to f^*E = X' \times_X E$ must send e to $(\tilde{\pi}(e), g(e))$ for the diagrams to commute.

Exercise 6.2.7   This follows abstractly from the discussed properties of the induced bundle, but explicitly f^*g is given by $1 \times f : Y \times_X E \to Y \times_X F$.

Exercise 6.2.8   By Exercise 6.2.7 it suffices to consider the trivial case of the product bundle, but the general case is not much harder: if $h: E \to X \times \mathbf{R}^n$ is a trivialization, then the map $f^*E = Y \times_X E \to Y \times_X (X \times \mathbf{R}^n)$ induced by h is a trivialization, since $Y \times_X (X \times \mathbf{R}^n) \to Y \times \mathbf{R}^n$ sending $(y, (x, v))$ to (y, v) is a homeomorphism.

Exercise 6.2.9
$$X \times_Y (Y \times_Z E) \cong X \times_Z E.$$

Exercise 6.2.10   The map
$$E \setminus \sigma_0(X) \to \pi_0^*E = (E \setminus \sigma_0(X)) \times_X E$$
sending v to (v, v) is a nonvanishing section.

Exercise 6.3.4   The transition functions will be of the type $U \to \mathrm{GL}_{n_1+n_2}(\mathbf{R})$, which sends $p \in U$ to the block matrix
$$\begin{bmatrix} (h_1)_p(g_1)_p^{-1} & 0 \\ 0 & (h_2)_p(g_2)_p^{-1} \end{bmatrix},$$
which is smooth if each of the blocks is smooth. More precisely, the transition function is a composite of three smooth maps:

(1) the diagonal $U \to U \times U$,
(2) the product
$$U \times U \longrightarrow \mathrm{GL}_{n_1}(\mathbf{R}) \times \mathrm{GL}_{n_2}(\mathbf{R})$$

   of the transition functions, and
(3) the block sum
$$(A, B) \mapsto \begin{bmatrix} A & 0 \\ 0 & B \end{bmatrix},$$

   namely
$$\mathrm{GL}_{n_1}(\mathbf{R}) \times \mathrm{GL}_{n_2}(\mathbf{R}) \to \mathrm{GL}_{n_1+n_2}(\mathbf{R}).$$

Similarly for the morphisms.

Exercise 6.3.6   Use the map $\epsilon \to S^n \times \mathbf{R}$ sending $(p, \lambda p)$ to (p, λ).

Exercise 6.3.8   Consider $TS^n \oplus \epsilon$, where ϵ is gotten from Exercise 6.3.6. Construct a trivialization $TS^n \oplus \epsilon \to S^n \times \mathbf{R}^{n+1}$.

Exercise 6.3.9

$$\epsilon_1 \oplus \epsilon_2 \xrightarrow[\cong]{h_1 \oplus h_2} X \times (\mathbf{R}^{n_1} \oplus \mathbf{R}^{n_2})$$

and

$$(E_1 \oplus E_2) \oplus (\epsilon_1 \oplus \epsilon_2) \cong (E_1 \oplus \epsilon_1) \oplus (E_2 \oplus \epsilon_2).$$

Exercise 6.3.10   Given f_1 and f_2, let $f: E_1 \oplus E_2 \to E_3$ be given by sending $(v, w) \in \pi_1^{-1}(p) \oplus \pi_2^{-1}(p)$ to $f_1(v) + f_2(w) \in \pi_3^{-1}(p)$. Given f, let $f_1(v) = f(v, 0)$ and $f_2(w) = f(0, w)$.

Exercise 6.4.4   Send the bundle morphism f to the section which to any $p \in X$ assigns the linear map $f_p: E_p \to E'_p$.

Exercise 6.4.5   For the bundle morphisms, you need to extend the discussion in Example 6.4.3 slightly and consider the map $\mathrm{Hom}(V_1, V_2) \times \mathrm{Hom}(V_3, V_4) \to \mathrm{Hom}(\mathrm{Hom}(V_2, V_3), \mathrm{Hom}(V_1, V_4))$ obtained by composition.

Exercise 6.4.6   The isomorphism $\mathbf{R} \cong \mathrm{Hom}(L_p, L_p)$ given by sending $a \in \mathbf{R}$ to multiplication by a extends to an isomorphism $X \times \mathbf{R} \cong \mathrm{Hom}(L, L)$. You will need to use the fact that multiplication in $\mathbf{R}$ is commutative.

As a matter of fact, by commutativity, the obvious atlas for $\mathrm{Hom}(L, L) \to X$ has trivial transition functions: if (g, U) and (h, V) are two bundle charts for $L \to X$, then the corresponding transition function for $\mathrm{Hom}(L, L) \to X$ is the map $U \cap V \to \mathrm{GL}(\mathrm{Hom}(\mathbf{R}, \mathbf{R}))$ sending p to $\mathrm{Hom}(h_p g_p^{-1}, g_p h_p^{-1}) = \{f \mapsto g_p h_p^{-1} f h_p g_p^{-1} = f\}$ (i.e., the identity element, regardless of p).

Exercise 6.4.7   Let $F \subseteq E$ be a rank-k subbundle of the rank-n vector bundle $\pi: E \to X$. Define as a set

$$E/F = \coprod_{p \in X} E_p/F_p$$

with the obvious projection $\bar{\pi}: E/F \to X$. The bundle atlas is given as follows. For $p \in X$, choose a bundle chart $h: \pi^{-1}(U) \to U \times \mathbf{R}^n$ such that $h(\pi^{-1}(U) \cap F) = U \times \mathbf{R}^k \times \{0\}$. On each fiber this gives a linear map on the quotient $\bar{h}_p: E_p/F_p \to \mathbf{R}^n/\mathbf{R}^k \times \{0\}$ via the formula $\bar{h}_p(\bar{v}) = \overline{h_p(v)}$ as in Section 6.4.1(2). This gives a function

$$\bar{h}: (\bar{\pi})^{-1}(U) = \coprod_{p \in U} E_p/F_p$$

$$\to \coprod_{p \in U} \mathbf{R}^n/\mathbf{R}^k \times \{0\}$$

$$\cong U \times \mathbf{R}^n/\mathbf{R}^k \times \{0\}$$

$$\cong U \times \mathbf{R}^{n-k}.$$

You then have to check that the transition functions $p \mapsto \bar{g}_p \bar{h}_p^{-1} = \overline{g_p h_p^{-1}}$ are continuous (or smooth).

Insofar as the map of quotient bundles is concerned, this follows similarly: define it on each fiber and check continuity of "up, over and down".

Exercise 6.4.9 Just write out the definition.

Exercise 6.4.10 First recall the local trivializations we used to define the tangent and cotangent bundles. Given a chart (x, U) for M, we have trivializations

$$(TM)_U \cong U \times \mathbf{R}^n$$

sending $[\gamma] \in T_p M$ to $(\gamma(0), (x\gamma)'(0))$ and

$$(T^* M)_U \cong U \times (\mathbf{R}^n)^*$$

sending $d\phi \in T_p^* M$ to $(p, D(\phi x^{-1})(x(p)) \cdot) \in U \times (\mathbf{R}^n)^*$. The bundle chart for the tangent bundle has inverse

$$U \times \mathbf{R}^n \cong (TM)_U$$

given by sending (p, v) to $[t \mapsto x^{-1}(x(p) + vt)] \in T_p M$, which gives rise to the bundle chart

$$(TM)_U^* \cong U \times (\mathbf{R}^n)^*$$

on the dual, sending $f \in (T_p M)^*$ to

$$(p, v) \mapsto f([t \mapsto x^{-1}(x(p) + vt)]).$$

The exercise is (more than) done if we show that the diagram

$$
\begin{array}{ccc}
(T^* M)_U & \xrightarrow{d\phi \mapsto \{[\gamma] \mapsto (\phi\gamma)'(0)\}} & (TM)_U^* \\
\cong \downarrow & & \cong \downarrow \\
U \times (\mathbf{R}^n)^* & = & U \times (\mathbf{R}^n)^*
\end{array}
$$

commutes, which it does since, if we start with $d\phi \in T_p^* M$ in the upper left-hand corner, we end up with $D(\phi x^{-1})(x(p)) \cdot$ either way (check that the derivative at $t = 0$ of $\phi x^{-1}(x(p) + vt)$ is $D(\phi x^{-1})(x(p)) \cdot v$).

Exercise 6.4.11 The procedure is just as for the other cases. Let $SB(E) = \coprod_{p \in X} SB(E_p)$. If (h, U) is a bundle chart for $E \to X$ define a bundle chart $SB(E)_U \to U \times SB(\mathbf{R}^k) \cong U \times \mathbf{R}^{k(k-n)/2}$ by means of the composite

$$
\begin{array}{ccc}
SB(E)_U & = & \coprod_{p \in U} SB(E_p) \\
& & \coprod SB(h_p^{-1}) \downarrow \\
U \times SB(\mathbf{R}^k) & = & \coprod_{p \in U} SB(\mathbf{R}^k).
\end{array}
$$

Explicitly, the transition function takes the form

$$U \to \mathrm{GL}_k \mathbf{R} \to \mathrm{GL}(SB(\mathbf{R}^k)),$$

where the first map is the corresponding transition function for E and the last map sends an invertible matrix A to the linear isomorphism sending a symmetric bilinear $f : \mathbf{R}^k \times \mathbf{R}^k \to \mathbf{R}$ to the composite

$$\mathbf{R}^k \times \mathbf{R}^k \xrightarrow{A \times A} \mathbf{R}^k \times \mathbf{R}^k \xrightarrow{f} \mathbf{R},$$

which depends polynomially, and hence smoothly, on the entries in A (to be totally sure, write it out in a basis for $SB(\mathbf{R}^k)$).

Exercise 6.4.13

$\mathrm{Alt}^k(E) = \coprod_{p \in X} \mathrm{Alt}^k E_p$ and so on.

Exercise 6.4.14

The transition functions on $L \to M$ are maps into nonzero real numbers, and on the tensor product this number is squared, and so all transition functions on $L \otimes L \to M$ map into positive real numbers.

Exercise 6.5.5

Prove that the diagonal $M \to M \times M$ is an imbedding by proving that it is an immersion inducing a homeomorphism onto its image. The tangent space of the diagonal at (p, p) is exactly the diagonal of $T_{(p,p)}(M \times M) \cong T_p M \times T_p M$. For any vector space V, the quotient space $V \times V/\text{diagonal}$ is canonically isomorphic to V via the map given by sending $(v_1, v_2) \in V \times V$ to $v_1 - v_2 \in V$.

Exercise 6.5.6

Consider the tautological line bundle $\eta_1 \to S^1$ as the subbundle of the trivial bundle given by

$$\{(e^{i\theta}, te^{i\theta/2}) \in S^1 \times \mathbf{C} \mid t \in \mathbf{R}\} \subseteq S^1 \times \mathbf{C}$$

as in Example 6.1.3. Then $\eta_1 \to \eta_1^\perp$ sending $(p, q) \to (p, iq)$ is an isomorphism, and so we get an isomorphism $\eta_1 \oplus \eta_1 \cong \eta_1 \oplus \eta_1^\perp \cong S^1 \times \mathbf{C}$.

Exercise 6.6.11

It is easiest to do this in the angle charts and plot the vectors displayed in Example 6.6.10. Different choices will give different "windows" as in the original robot example in Section 1.1.

Exercise 6.7.1

The conditions you need are exactly the ones fulfilled by the elementary definition of the determinant: check your freshman introduction.

Exercise 6.8.1

Check out, e.g., page 59 of [16].

Exercise 6.8.2

Check out, e.g., page 60 of [16].

Chapter 7

Exercise 7.1.4

Check the two defining properties of a flow. As an aside: this flow could be thought of as the flow $\mathbf{R} \times \mathbf{C} \to \mathbf{C}$ sending (t, z) to $z^{-1}e^{-t/2}$, which obviously satisfies the two conditions.

Exercise 7.1.9

Symmetry ($\Phi(0, p) = p$) and reflexivity ($\Phi(-t, \Phi(t, p)) = p$) are obvious, and transitivity follows since if

$$p_{i+1} = \Phi(t_i, p_i), \qquad i = 0, 1$$

then

$$p_2 = \Phi(t_1, p_1) = \Phi(t_1, \Phi(t_0, p_0)) = \Phi(t_1 + t_0, p_0).$$

Exercise 7.1.11

(i) Flow lines are constant. (ii) All flow lines outside the origin are circles. (iii) All flow lines outside the origin are rays flowing towards the origin.

Exercise 7.1.17

Note that, since Φ_s is a diffeomorphism and $T\Phi_s \dot{\phi}_p(0) = T\Phi_s[\phi_p] = \dot{\phi}_p(s)$, the velocity vector $\dot{\phi}_p(s)$ is either zero for all s or never zero at all.

If $\dot{\phi}_p(s) = 0$ for all s, this means that ϕ_p is constant since if (x, U) is a chart with $\phi_p(s_0) \in U$ we get that $(x\phi_p)'(s) = 0$ for all s close to s_0, hence $x\phi_p(s)$ is constant for all s close to s_0 giving that ϕ_p is constant.

If $\dot{\phi}_p(s) = T\phi_p[L_s]$ is never zero we get that $T\phi_p$ is injective ($[L_s] \neq 0 \in T_s\mathbf{R} \cong \mathbf{R}$), and so ϕ_p is an immersion. To conclude we must treat the case where ϕ_p is not injective, i.e., there are numbers $s < s'$ with $\phi_p(s) = \phi_p(s')$. This means that

$$\begin{aligned}
p = \phi_p(0) = \Phi(0, p) &= \Phi(-s + s, p) \\
&= \Phi(-s, \Phi(s, p)) = \Phi(-s, \phi_p(s)) \\
&= \Phi(-s, \phi_p(s')) = \Phi(-s + s', p) \\
&= \phi_p(-s + s').
\end{aligned}$$

Since ϕ_p is continuous $\phi_p^{-1}(p) \subseteq \mathbf{R}$ is closed and not empty (it contains 0 and $-s+s' > 0$ among others). As ϕ_p is an immersion it is a local imbedding, so there is an $\epsilon > 0$ such that

$$(-\epsilon, \epsilon) \cap \phi_p^{-1}(p) = \{0\}.$$

Hence the set

$$S = \{t > 0 \mid p = \phi_p(t)\} = \{t \geq \epsilon \mid p = \phi_p(t)\}$$

is closed and bounded below. This means that there is a smallest positive number T such that $\phi_p(0) = \phi_p(T)$. Clearly $\phi_p(t) = \phi_p(t + kT)$ for all $t \in \mathbf{R}$ and any integer k.

On the other hand, we get that $\phi_p(t) = \phi_p(t')$ **only if** $t - t' = kT$ for some integer k. For, if $(k-1)T < t-t' < kT$, then $\phi_p(0) = \phi_p(kT - (t-t'))$ with $0 < kT - (t-t') < T$, contradicting the minimality of T.

Exercise 7.1.20
Consider one of the "bad" injective immersions that fail to be imbeddings, and force a discontinuity on the velocity field.

Exercise 7.2.3
Consider a bump function ϕ on the sphere which is 1 near the North pole and 0 near the South pole. Consider the vector field $\vec{\Phi} = \phi\vec{\Phi}_N + (1 - \phi)\vec{\Phi}_S$. Near the North pole $\vec{\Phi} = \vec{\Phi}_N$ and near the South pole $\vec{\Phi} = \vec{\Phi}_S$, and so the flow associated with $\vec{\Phi}$ has the desired properties (that t is required to be small ensures that we do not flow from one pole to another).

Exercise 7.2.4
The vector field associated with the flow
$$\Phi \colon \mathbf{R} \times (S^1 \times S^1) \to (S^1 \times S^1)$$
given by $\Phi(t, (z_1, z_2)) = (e^{iat}z_1, e^{ibt}z_2)$ exhibits the desired phenomena on varying the real numbers a and b.

Exercise 7.2.5
All we have to show is that X is the velocity field of Φ. Under the diffeomorphism
$$TO(n) \to E,$$
$$[\gamma] \to (\gamma(0), \gamma'(0))$$
this corresponds to the observation that
$$\frac{\partial}{\partial s}\bigg|_{s=0} \Phi(s, g) = gA.$$

Exercise 7.3.10
Do a variation of Example 7.3.4.

Exercise 7.4.3
There is no hint other than use the definitions!

Exercise 7.4.4
Use the preceding exercise: notice that $T\pi_M\xi = \pi_{TM}\xi$ is necessary for things to make sense since $\ddot{\gamma}$ had two repeated coordinates.

Exercise 7.4.7
Insofar as the isomorphism $F \cong TTS^n$ is concerned, note that a curve $t \mapsto (p(t), v(t))$ in E must satisfy $p(t) \cdot p(t) = 1$ and $p(t) \cdot v(t) = 0$, which upon differentiation gives $2p(t) \cdot p'(t) = p'(t) \cdot v(t) + p(t) \cdot v'(t) = 0$. Evaluating at $t = 0$ gives exactly $|p(0)| = 1$, $p(0) \cdot v(0) = p(0) \cdot p'(0) = p'(0) \cdot v(0) + p(0) \cdot v'(0) = 0$. The curve γ is a solution curve to ξ since $\gamma(0) = p$, $\gamma'(0) = v$ and $\gamma''(t) = -|v|^2\gamma(t)$.

Exercise 7.4.8

It suffices to check $(T\pi_M)\xi = \mathrm{id}_{TM}$ locally, so we may assume $M = \mathbf{R}^n$. Under the identifications $T\mathbf{R}^n \cong \mathbf{R}^n \times \mathbf{R}^n$ and $TT\mathbf{R}^n \cong \mathbf{R}^n \times \mathbf{R}^n \times \mathbf{R}^n \times \mathbf{R}^n$ of Exercise 7.4.3 the second-order differential equation ξ_i corresponds to $(p, v) \mapsto (p, v, v, f_i(p, v))$ for $i = 1, 2$, and $s_1\xi_1 + s_2\xi_2$ to

$$(p, v) \mapsto s_1(p, v, v, f_1(p, v)) + s_2(p, v, v, f_2(p, v))$$
$$= (p, v, s_1 v + s_2 v, s_1 f_1(p, v) + s_2 f_2(p, v))$$
$$= (p, v, v, s_1 f_1(p, v) + s_2 f_2(p, v)).$$

Chapter 8

Exercise 8.2.3

Consider a partition of unity $\{\phi_i\}_{i \in \mathbf{N}}$ as displayed in the proof of Theorem 8.2.2 where $\mathrm{supp}(\phi) = x_i^{-1} E^n(2)$ for a chart (x_i, U_i) for each i. Let f_i be the composite

$$E^n(2) \xrightarrow{\cong} x_i^{-1}(E^n(2)) = \mathrm{supp}(\phi_i) \xrightarrow{f|_{\mathrm{supp}(\phi_i)}} \mathbf{R}$$

and choose a polynomial g_i such that $|f_i(x) - g_i(x)| < \epsilon$ for all $x \in E^n(2)$. Let $g(p) = \sum_i \phi_i(p) g_i(x_i(p))$, which gives a well-defined and smooth map. Then

$$|f(p) - g(p)| = |\sum_i \phi_i(p)(f(p) - g_i(x_i(p)))|$$

$$= |\sum_i \phi_i(p)(f_i(x_i(p)) - g_i(x_i(p)))|$$

$$\leq \sum_i \phi_i(p)|f_i(x_i(p)) - g_i(x_i(p))|$$

$$< \sum_i \phi_i(p)\epsilon = \epsilon.$$

Exercise 8.2.4

By Exercise 6.4.14, all the transition functions $U \cap V \to \mathrm{GL}_1(\mathbf{R})$ in the associated bundle atlas on the line bundle $L \otimes L \to M$ have values in the positive real numbers. So let us prove the following more general statement.

Lemma *Let $L \to M$ be a smooth line bundle with a smooth atlas such that all transition functions have positive values. Then $L \to M$ is trivial.*

Proof. Choose a partition of unity $\{\phi_i\}_{i \in \mathbf{N}}$ subordinate to this atlas. For each $i \in \mathbf{N}$, choose a bundle chart (h^i, U^i) such that $\mathrm{supp}\,\phi_i \subseteq U^i$. Define $h \colon L \to M \times \mathbf{R}$ by $h(p, e) = (p, \sum_i \phi_i(p) h_p^i(e))$. We check whether (h, M) is a bundle chart in the maximal atlas by checking that, for each j, the transition function $h_{U^j}(h^j)^{-1}$ is smooth and a linear isomorphism on each fiber. For $p \in U^j$,

$$h_p(h_p^j)^{-1} = \sum_i \phi_i(p) h_p^i(h_p^j)^{-1},$$

so smoothness and linearity are assured. Since $h_p^i(h_p^j)^{-1}$ corresponds to multiplying by a positive number, this convex combination does too, and so $h_{U^j}(h^j)^{-1}$ is an isomorphism on each fiber. Hence $L \to M$ is trivial. $\blacksquare$

Exercise 8.2.8

In the space of all vector fields on TM, the property of being a second-order differential equation is convex (see Exercise 7.4.8) and the spray property is linear. To see the linearity of the spray condition it suffices to implement it locally and assume $M = \mathbf{R}^n$. Under the identifications $T\mathbf{R}^n \cong \mathbf{R}^n \times \mathbf{R}^n$ and $TT\mathbf{R}^n \cong \mathbf{R}^n \times \mathbf{R}^n \times \mathbf{R}^n \times \mathbf{R}^n$ of Exercise 7.4.3 the spray condition on a vector field $\xi: TM \to TTM$ corresponds to saying that the associated function

$$(p, v) \mapsto (p, v, f_1(p, v), f_2(p, v))$$

must satisfy $f_1(p, sv) = sf_1(p, v)$ and $f_2(p, sv) = s^2 f_2(p, v)$ for all $s \in \mathbf{R}$. If $(p, v) \mapsto (p, v, g_1(p, v), g_2(p, v))$ represents another such spray and $s_1, s_2 \in \mathbf{R}$, then $s_1 f_i(p, sv) + s_2 g_i(p, sv) = s^i(s_1 f_i(p, v) + s_2 g_i(p, v))$ for $i = 1, 2$.

All in all, the conditions the sprays have to satisfy are convex and so can be glued together (c.f. the proof of existence of fiber metrics Theorem 8.3.1 in the next section). Choose a countable atlas $\{(x_i, U_i)\}$ and a partition of unity $\{\phi_i\}$ subordinate to $\{U_i\}$. For each i, choose any spray $\xi_i: TU_i \to TTU_i$ (e.g., the one associated with $(p, v) \mapsto (p, v, v, 0)$) and let

$$\xi = \sum_i \phi_i \xi_i : TM \to TTM,$$

which by convexity is a spray.

Exercise 8.2.10

Let us do it directly from the definitions! Recall the vector field $\vec{L}: \mathbf{R} \to T\mathbf{R}$ given by $\vec{L}(r) = [t \mapsto r + t]$. Observe that $Ts\,\vec{L}(r) = [t \mapsto sr + st] = s\,\vec{L}(sr)$. Then $\dot{\gamma}_s(r) = T\gamma_s \vec{L}(r) = T\gamma Ts\vec{L}(r) = T\gamma s\vec{L}(sr) = s\,T\gamma\vec{L}(sr) = s\dot{\gamma}(sr)$, so that $\dot{\gamma}_s(0) = s\dot{\gamma}(0)$. Finally, $\ddot{\gamma}_s = T\dot{\gamma}_s\vec{L} = T(sT\gamma\vec{L})\vec{L} = Ts\,TT\gamma\,T\vec{L}\,Ts\vec{L} = Ts\,TT\gamma\,T\vec{L}s\vec{L} = Tss\,TT\gamma\,T\vec{L}\,\vec{L}s = Tss\,\ddot{\gamma}s = Tss\xi\dot{\gamma}s = \xi s\dot{\gamma}s = \xi\dot{\gamma}_s$.

Exercise 8.2.11

The thing to check is that $\mathcal{T}$ is an open neighborhood of the zero section.

Exercise 8.2.12

In the notation introduced in Exercise 7.4.7, if $s \in \mathbf{R}$, the map $s: TS^n \to TS^n$ corresponds to the map $E \to E$ sending (p, v) to (p, sv), and the maps $s, Ts: TTS^n \to TTS^n$ correspond to the maps $F \to F$ sending (p, v_1, v_2, v_3) to (p, v_1, sv_2, sv_3) and (p, sv_1, v_2, sv_3). Since ξ corresponds to the map $E \to F$ sending (p, v) to $(p, v, v, -|v|^2 p)$, both ξs and $Ts\,(s\xi)$ correspond to sending (p, v) to $(p, sv, sv, s^2(-|v|^2 p))$.

The formula for the exponential now follows by flowing along the solution curve for a time $t = 1$.

Exercise 8.3.5

Use Lemma 6.6.13 to show that the bundle in question is isomorphic to $(T\mathbf{R}^n)|_M \to M$.

Exercise 8.3.6

You have done this exercise before!

Exercise 8.3.7

Analyzing $\mathrm{Hom}(\eta_n, \eta_n^\perp)$ we see that we may identify it with the set of pairs $(L, \alpha\colon L \to L^\perp)$, where $L \in \mathbf{RP}^n$ and α a linear map. On the other hand, Exercise 5.5.19 identifies $T\mathbf{RP}^n$ with $\{(p, v) \in S^n \times \mathbf{R}^{n+1} \mid p \cdot v = 0\}/(p, v) \sim (-p, -v)$. This means that we may consider the bijection $\mathrm{Hom}(\eta_n, \eta_n^\perp) \to T\mathbf{RP}^n$ given by sending $(L, \alpha\colon L \to L^\perp)$ to $\pm(p, \alpha(p))$, where $\pm p = L \cap S^n$. This bijection is linear on each fiber. Check that it defines a bundle morphism by considering trivializations over the standard atlas for $\mathbf{RP}^n$.

Exercise 8.3.8

By Exercise 6.4.6 the Hom-bundle $\mathrm{Hom}(\eta_n, \eta_n) \to \mathbf{RP}^n$ is trivial, so by Exercise 6.4.5 we have an isomorphism

$$T\mathbf{RP}^n \oplus \epsilon \cong \mathrm{Hom}(\eta_n, \eta_n^\perp) \oplus \mathrm{Hom}(\eta_n, \eta_n).$$

By the natural isomorphism

$$\mathrm{Hom}(E, F) \oplus \mathrm{Hom}(E, F') \cong \mathrm{Hom}(E, F \oplus F'),$$

the latter bundle is isomorphic to $\mathrm{Hom}(\eta_n, \eta_n^\perp \oplus \eta_n)$, which is isomorphic to

$$\mathrm{Hom}(\eta_n, \epsilon \oplus \cdots \oplus \epsilon)$$

since $\eta_n^\perp \oplus \eta_n$ is trivial. By the same argument, we get an isomorphism to

$$\mathrm{Hom}(\eta_n, \epsilon \oplus \cdots \oplus \epsilon) = \eta_n^* \oplus \cdots \oplus \eta_n^*.$$

Now, choosing a fiber metric, we get an isomorphism $\eta_n^* \cong \eta_n$ and we are done.

Insofar as the last question is concerned, observe that $\mathbf{RP}^3 \cong SO(3)$ is parallelizable and that the induced bundle of a trivial bundle is trivial.

Exercise 8.3.9

The module structure $C^\infty(M) \times \Gamma(E) \to \Gamma(E)$ sends (f, σ) to the section $p \mapsto f(p) \cdot \sigma(p)$ (scalar multiplication in the fiber). If $E \to M$ is a subbundle of a product bundle $T = M \times R^N \to M$, observe that we have an isomorphism of vector bundles $T \cong E \oplus T/E$ and so a $C^\infty(M)$-isomorphism $C^\infty(M)^{\times N} \cong \Gamma(T) \cong \Gamma(E) \oplus \Gamma(T/E)$, i.e., $\Gamma(E)$ is a direct summand of a $C^\infty(M)$-free module (which is one characterization of being projective).

Exercise 8.4.9

Given a positive function f, it is equivalent that f and its square $g = f \cdot f$ are Morse, and if they are Morse the indices agree: the Hessian of g will contain both first and second derivatives of f, but at a critical point p the first derivatives vanish and all that remains is $2f(p)$ (which is a positive number) times the Hessian of f at p.

Using angle charts g is given by $g(e^{i\theta}, e^{i\phi}) = 11 - 6\cos\theta - 6\cos\phi + 2\cos(\theta - \phi)$ and the Hessian is

$$\begin{bmatrix} 6\cos\theta - 2\cos(\theta - \phi) & 2\cos(\theta - \phi) \\ 2\cos(\theta - \phi) & 6\cos\phi - 2\cos(\theta - \phi) \end{bmatrix}.$$

Upon inserting the critical points (i.e., (θ, ϕ) equal to $(0, 0)$, (π, π), $(0, \pi)$ and $(\pi, 0)$ with corresponding critical values 1, 5, 3 and 3) we get the matrices

$$\begin{bmatrix} 4 & 2 \\ 2 & 4 \end{bmatrix}, \begin{bmatrix} -8 & 2 \\ 2 & -8 \end{bmatrix}, \begin{bmatrix} 8 & -2 \\ -2 & -8 \end{bmatrix}, \begin{bmatrix} -4 & -2 \\ -2 & 4 \end{bmatrix}.$$

The numbers of negative eigenvalues of these matrices (and hence the indices of g) are 0, 2, 1 and 1, respectively.

Exercise 8.4.10

We have $f'(t) = 4t^3 - 2at$ and $f''(t) = 12t^2 - 2a$. The critical points are $t = 0$, and possibly $\pm\sqrt{a/2}$, which we insert into f'' and get $-2a$ and (possibly, twice) $4a$. Hence, f is Morse iff $a \neq 0$.

Exercise 8.4.13

The charts around interior points are inherited from M by intersection. At the boundary Theorem 4.4.3 guarantees us charts in M where $f^{-1}(a)$ is cut out by $R^{n-1} \times \{0\} \subseteq R^n$. By continuity M^a will occupy either the upper or the lower half space.

Exercise 8.5.3

If $s \colon M \to E$ is a nonvanishing vector field, then $m \mapsto s(m)/|s(m)|$ is a section of $S(E) \to M$.

Exercise 8.5.4

Let $i \colon S^{n-1} \subseteq R^n$ be the inclusion and let (x, U) be shorthand for the chart $(x^{0,0}, U^{0,0})$ in the standard atlas on S^{n-1} (given by the projection to the last $n-1$ coordinates). If $\phi \colon S^{n-1} \to S^{n-1}$ is a diffeomorphism, then the Jacobi matrix $D(i\phi x^{-1})(0) \in M_{n \times (n-1)}R$ has linearly independent columns lying in the orthogonal complement of $i\phi x^{-1}(0)$. Hence, the $n \times n$ matrix $M(\phi) = [i\phi x^{-1}(0) \; D(i\phi x^{-1})(0)]$ is invertible. Also, note that, if $A \in O(n)$ is considered as a diffeomorphism of the sphere, then the chain rule (and the fact that the Jacobi matrix $DA(p) = A$ is independent of p) gives that $M(A\phi) = A \cdot M(\phi)$.

Choose a bundle atlas $\mathcal{A}$ for $\pi \colon E \to M$ with all transition functions mapping to $O(n)$. Let $(h, U) \in \mathcal{A}$, and define $h' \colon E_U \to U \times R^n$ as $h'(p, v) = (p, M(h_p f_p^{-1})^{-1} h_p v)$. Since $h'_p h_p^{-1} = M(h_p f_p^{-1})^{-1} \in GL_n R$ depends smoothly on p (check this!), (h', U) is a bundle chart in the maximal smooth bundle atlas. Given another $(g, V) \in \mathcal{A}$, the transition function for (h', U) and (g', V) is

$$p \mapsto g'_p (h'_p)^{-1} = M(g_p f_p^{-1})^{-1} g_p h_p^{-1} M(h_p f_p^{-1})$$
$$= M(g_p f_p^{-1})^{-1} M(g_p h_p^{-1} h_p f_p^{-1}) = \mathrm{id}.$$

Hence, $\{(h', U)\}_{(h,U) \in \mathcal{A}}$ is a smooth bundle atlas whose transition functions are constant and equal to the identity matrix, proving that our bundle is trivial.

Exercise 8.5.5

Let $\pi \colon E \to M$ be a locally trivial smooth fibration with M a connected nonempty smooth manifold. Choose a $p \in M$ and let $F = \pi^{-1}(p)$. Consider the set

$$U = \{x \in M \mid \pi^{-1}(x) \cong F\},$$

and let V be the complement. We will show that both U and V are open, and so $U = M$ since $p \in U$ and M is connected. If $x \in U$ choose a trivializing neighborhood $x \in W$,

$$h \colon \pi^{-1}(W) \to W \times \pi^{-1}(x).$$

Now, if $y \in W$, then h induces a diffeomorphism between $\pi^{-1}(y)$ and $\pi^{-1}(x) \cong F$, so U is open. Likewise for V.

Exercise 8.5.7

Let $f \colon E \to M$ be a locally trivial fibration with compact fibers. Assume M is connected and let $K \subseteq M$ be compact. Let $(x_1, U_1), \ldots, (x_k, U_k)$ be charts from a good atlas such that $K \subseteq \bigcup_{i=1}^k E_i$, where $E_i = x_i^{-1} E^n(1)$, and such that f is trivial when restricted to U_i. Since $f^{-1}K \subseteq \bigcup_{i=1}^k f^{-1}\bar{E}_i$ it suffices to observe that each $f^{-1}\bar{E}_i \cong \bar{E}_i \times f^{-1}(x_i^{-1}(0))$ is compact.

Exercise 8.5.8

If $K \subseteq Z$ is compact, then $(gf)^{-1}(K) = f^{-1}g^{-1}(K)$, so, if both f and g are proper, so is gf. If $K \subseteq Y$ is compact, then $g(K) \subseteq Z$ is compact, and if gf is proper, so is $(gf)^{-1}g(K) \subseteq X$. But $f^{-1}(K) \subseteq (gf)^{-1}g(K)$ is a closed subspace, and hence is compact.

Exercise 8.5.9

This follows from the corresponding statements for surjections.

Exercise 8.5.14

Since $f\Phi_i(t, q') = f(q') + e_i t$ for all $q' \in E$ we get that $f\phi(t, q) = f(q) + t = r_0 + t$ for $q \in f^{-1}(r_0)$. This gives that the first coordinate of $\phi^{-1}\phi(t, q)$ is t, and that the second coordinate is q follows since $\Phi_i(-t_i, \Phi_i(t_i, q')) = q'$. Similarly for the other composite.

Exercise 8.5.15

The inclusion $\mathbf{R} - \{0\} \subseteq \mathbf{R}$ is an example, but if you want a surjective example the projection to the first coordinate $\mathbf{R}^2 - \{0\} \to \mathbf{R}$ will do.

Exercise 8.5.16

Concerning the map $\ell \colon S^1 \to \mathbf{CP}^1$, note that it maps into a chart domain on which Lemma 8.5.12 tells us that the projection is trivial.

Exercise 8.5.17

Write $\mathbf{R}$ as a union of intervals J_j so that, for each j, $\gamma(U_j)$ is contained within one of the open subsets of M so that the fibration trivializes. On each of these intervals the curve lifts, and you may glue the liftings using bump functions.

Exercise 8.5.18

Since $O(n + 1)$ is compact, it suffices by Corollary 8.5.13 to show that f is a submersion. Recall from Exercise 5.5.20 that under the isomorphism

$$T M_n(\mathbf{R}) \cong M_n(\mathbf{R}) \times M_n(\mathbf{R}),$$
$$[\gamma] \leftrightarrows (\gamma(0), \gamma'(0))$$

we get an isomorphism

$$E = \{(g, A) \in O(n) \times M_n(\mathbf{R}) \mid A^{\mathrm{T}} = -g^{\mathrm{T}} A g^{\mathrm{T}}\} \cong T O(n).$$

If we also use the identification

$$T S^n \cong \{(p, v) \in S^n \times \mathbf{R}^{n+1} \mid v^{\mathrm{T}} p = 0\},$$

we get that the map f sends $(g, A) \in E$ to $(g \cdot e_1, A \cdot e_1)$. So, we must show that, if $v \in \mathbf{R}^{n+1}$ and $v^{\mathrm{T}}(ge_1) = 0$, then there is an $A \in M_{n+1}\mathbf{R}$ with $A^{\mathrm{T}} = -g^{\mathrm{T}} A g^{\mathrm{T}}$ and $Ae_1 = v$. Let $B = g^{\mathrm{T}} A$ and $w = g^{\mathrm{T}} v$. Then the demands translate to

(1) $w^{\mathrm{T}} e_1 = 0$ (first entry of w is zero),
(2) $B = -B^{\mathrm{T}}$ (B is skew symmetric), and
(3) $Be_1 = w$ (first column of B is w),

which are satisfied by the matrix whose first column is w, whose top row is $-w^{\mathrm{T}}$ (which is OK since the first entry is zero), and which is otherwise zero.

Since S^n is connected for $n > 0$ the fibers are all diffeomorphic, so it suffices to check the fiber over e_1, which consists of matrices $A \in O(n + 1)$ with first column e_1. Hence A is uniquely given by the lower-right $n \times n$ submatrix which is necessarily orthogonal too. When $n = 1$ we have that $O(1) \to S^0$ is a bijection.

Exercise 8.5.19

Exercise 8.5.7 gives that the associated sphere bundle is a proper submersion, which is trivial by Lemma 8.5.12. Hence, Exercise 8.5.4 finishes the proof.

Exercise 8.5.20

Of course, one *could* use the Ehresmann fibration theorem and Exercise 4.7.12 in the case when π is proper by showing that then ϕ is proper too, but it is easier to just prove the local triviality directly. In fact, given a local trivialization $\pi^{-1}(U) \cong U \times F$ of π, one gets a trivialization $\phi^{-1}(f^{-1}(U)) \cong f^{-1}(U) \times_U \pi^{-1}(U) \cong f^{-1}(U) \times F$.

Exercise 8.5.21

Observe that $H(m, n)$ is compact, so by the compact version of the Ehresmann fibration theorem, Corollary 8.5.13, all we have to show is that $\pi: H(m, n) \to \mathbf{RP}^m$ is a submersion. Consider the submersion

$$f: (\mathbf{R}^{m+1} \setminus \{0\}) \times (\mathbf{R}^{n+1} \setminus \{0\}) \to \mathbf{R}$$

given by $f(([p], [q])) = \sum_{k=0}^m p_k q_k$, and consider the submanifold $N = f^{-1}(0)$. Now, if $\pi': N \to \mathbf{R}^{m+1} \setminus \{0\}$ is the projection given by $\pi'(p, q) = p$, then

$$
\begin{array}{ccc}
N & \xrightarrow{\pi'} & \mathbf{R}^{m+1} \setminus \{0\} \\
{\scriptstyle \text{projection}} \downarrow & & \downarrow {\scriptstyle \text{projection}} \\
H(m, n) & \xrightarrow{\pi} & \mathbf{RP}^m
\end{array}
$$

commutes, and so it suffices to show that π' is a submersion.

Lastly, since $\mathbf{RP}^m$ is connected, to show that all fibers are diffeomorphic to $\mathbf{RP}^{n-1}$ it suffices to see that $\pi^{-1}([1, 0, \ldots, 0]) = \{([1, 0, \ldots, 0], [0, q_1, \ldots, q_n])\}$ is diffeomorphic to $\mathbf{RP}^{n-1}$.

Exercise 8.5.22

Since $SO(n)$ is compact and connected we need show only that f is a submersion and that $f^{-1}(f(I)) \cong SO(n - k)$. To show that f is a submersion, let $A \in SO(n)$ and consider the maps $L_A: SO(n) \to SO(n)$ and $L_A: V_n^k \to V_n^k$ given by $L_A(B) = A \cdot B$. Show that both are diffeomorphisms and that $L_A f = f L_A$. Hence

$$
\begin{array}{ccc}
T_I SO(n) & \xrightarrow[\cong]{T_I L_A} & T_A SO(n) \\
{\scriptstyle T_I f} \downarrow & & \downarrow {\scriptstyle T_A f} \\
T_{f(I)} V_n^k & \xrightarrow[\cong]{L_A} & T_{f(A)} V_n^k
\end{array}
$$

commutes, and showing that f is a submersion reduces to showing that $T_I f$ is surjective. First, recall from Exercise 5.5.20 the isomorphism

$$T_I SO(n) \cong \mathrm{Skew}(n)$$
$$= \{X \in M_n(\mathbf{R}) \mid X^T = -X\}.$$

In the same manner, one establishes

$$T_{f(I)} V_n^k \cong \mathrm{Skew}(k, n) = \left\{ \begin{bmatrix} A \\ B \end{bmatrix} \in M_{n \times k}(\mathbf{R}) \mid A^T = -A \right\}$$

and that under these isomorphisms $T_I f$ corresponds to the surjection sending the skew matrix $\begin{bmatrix} A & -B^T \\ B & C \end{bmatrix}$ to $\begin{bmatrix} A \\ B \end{bmatrix}$.

Exercise 8.5.23

Just as in Exercise 8.5.22, this exercise reduces to showing that $T_I C$ is a surjection. Again we use the isomorphism

$$T_I SO(n) \cong Skew(n)$$

(the image of $T_I SO(n)$ in $T_I GL_n(\mathbf{R})$ is exactly $\{[s \mapsto I + sB] \in T_I GL_n(\mathbf{R}) \mid B^T = -B\}$). Recall from Example 2.3.15 the chart

$$x_{\mathbf{R}^k} : U_{\mathbf{R}^k} \to Hom(\mathbf{R}^k, (\mathbf{R}^k)^\perp)$$

sending $V = C(A) \in U_{\mathbf{R}^k}$ to the linear map

$$f_A = x_{\mathbf{R}^k}(V) : \mathbf{R}^k \to (\mathbf{R}^k)^\perp,$$

where $f_A(p) \in (\mathbf{R}^k)^\perp$ is such that $p + f_A(p) \in V$.

To show that I is a regular point it suffices to show that $Skew(n) \to T_0 Hom(\mathbf{R}^k, (\mathbf{R}^k)^\perp)$ sending B to the tangent vector $[s \mapsto f_{I+sB}]$ is surjective ($C(A)$ makes sense for any $A \in GL_n(\mathbf{R})$ and $I + sB$ is invertible when s is small). That is, given $X \in Hom(\mathbf{R}^k, (\mathbf{R}^k)^\perp) \cong M_{(n-k)\times k}(\mathbf{R})$, find $B \in Skew(n)$ such that $[s \mapsto f_{I+sB}] = [s \mapsto sX]$.

Choosing

$$B = \begin{bmatrix} 0 & -X^T \\ X & 0 \end{bmatrix}$$

does the trick: $f_{I+sB}(p) = sXp$.

Exercise 8.5.24

This follows by Exercise 8.5.9 since $C = Sf$ is a submersion by Exercise 8.5.23. Of course, we **could** have chosen to establish Exercise 8.5.24 from scratch and deduced Exercise 8.5.23 from Exercise 8.5.22 by saying that the composites of submersions are submersions.

Appendix A

Exercise A.1.3

Consider the union of the closed intervals $[1/n, 1]$ for $n \geq 1$.

Exercise A.1.5

Consider the set of all open subsets of X contained in U. Its union is open.

Exercise A.1.7

By the union axiom for open sets, int A is open and contains all open subsets of A.

Exercise A.1.8

The intersection of two open balls is the union of all open balls contained in the intersection.

Exercise A.1.9

All open intervals are open balls!

Exercise A.2.2

Hint for one way: the "existence of the δ" assures you that every point in the inverse image has a small interval around it inside the inverse image of the ϵ ball (Figure B.6).

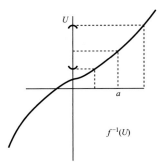

Figure B.6.

Exercise A.2.3 $f^{-1}(g^{-1}(U)) = (gf)^{-1}(U)$.

Exercise A.2.6   Use first-year calculus.

Exercise A.3.3   Can you prove that the set containing only the intervals (a, b) when a and b varies over the rational numbers is a basis for the usual topology on the real numbers?

Exercise A.3.4   Show that, given a point and an open ball containing the point, there is a "rational" ball in between.

Exercise A.3.5   Use Note A.3.2.

Exercise A.3.6 $f^{-1}(\bigcup_\alpha V_\alpha) = \bigcup_\alpha f^{-1}(V_\alpha)$.

Exercise A.5.2   Use that $\left(\bigcup_\alpha U_\alpha\right) \cap A = \bigcup_\alpha (U_\alpha \cap A)$ and $\left(\bigcap_\alpha U_\alpha\right) \cap A = \bigcap_\alpha (U_\alpha \cap A)$.

Exercise A.5.3   Use Exercise A.2.3 one way, and that if $f(Z) \subseteq A$, then $f^{-1}(U \cap A) = f^{-1}(U)$ the other.

Exercise A.5.4   The intersections of A with the basis elements of the topology on X will form a basis for the subspace topology on A.

Exercise A.5.5   Separate points in A by means of disjoint open neighborhoods in X, and intersect with A.

Exercise A.6.4   Inverse image commutes with union and intersection.

Exercise A.6.5 Use Exercise A.2.3 one way, and the characterization of open sets in $X/\sim$ for the other.

Exercise A.6.6 Show that open sets in one topology are open in the other.

Exercise A.7.2 Cover $f(X) \subseteq Y$ by open sets, i.e., by sets of the form $V \cap f(X)$ where V is open in Y. Since $f^{-1}(V \cap f(X)) = f^{-1}(V)$ is open in X, this gives an open cover of X. Choose a finite subcover, and select the associated Vs to cover $f(X)$.

Exercise A.7.5 The real projective space is compact by Exercise A.7.2. The rest of the claims follow by Theorem A.7.11, but you can give a direct proof by following the outline below.

For $p \in S^n$ let $[p]$ be the equivalence class of p considered as an element of $\mathbf{RP}^n$. Let $[p]$ and $[q]$ be two different points. Choose an ϵ such that ϵ is less than both $|p - q|/2$ and $|p + q|/2$. Then the ϵ balls around p and $-p$ do not intersect the ϵ balls around q and $-q$, and their images define disjoint open sets separating $[p]$ and $[q]$.

Notice that the projection $p\colon S^n \to \mathbf{RP}^n$ sends open sets to open sets, and that, if $V \subseteq \mathbf{RP}^n$, then $V = pp^{-1}(V)$. This implies that the countable basis on S^n inherited as a subspace of $\mathbf{R}^{n+1}$ maps to a countable basis for the topology on $\mathbf{RP}^n$.

Exercise A.7.9 You must show that, if $K \subseteq C$ is closed, then $\left(f^{-1}\right)^{-1}(K) = f(K)$ is closed.

Exercise A.7.10 Use Theorem A.7.3 (Heine–Borel) and Exercise A.7.2.

Exercise A.8.2 One way follows by Exercise A.2.3. For the other, observe that by Exercise A.3.6 it suffices to show that, if $U \subseteq X$ and $V \subseteq Y$ are open sets, then the inverse image of $U \times V$ is open in Z.

Exercise A.8.3 Show that a square around any point contains a circle around the point and vice versa.

Exercise A.8.4 If $\mathcal{B}$ is a basis for the topology on X and $\mathcal{C}$ is a basis for the topology on Y, then

$$\{U \times V \mid U \in \mathcal{B}, V \in \mathcal{C}\}$$

is a basis for $X \times Y$.

Exercise A.8.5 If $(p_1, q_1) \neq (p_2, q_2) \in X \times Y$, then either $p_1 \neq p_2$ or $q_1 \neq q_2$. Assume the former, and let U_1 and U_2 be two open sets in X separating p_1 and p_2. Then $U_1 \times Y$ and $U_2 \times Y$ are $\ldots$

Exercise A.9.2 The inverse image of a set that is both open and closed is both open and closed.

Exercise A.9.5   Both X_1 and X_2 are open sets.

Exercise A.9.6   One way follows by Exercise A.2.3. The other follows since an open subset of $X_1 \coprod X_2$ is the (disjoint) union of an open subset of X_1 with an open subset of X_2.

Exercise A.10.4   If $p \in f(f^{-1}(B))$ then $p = f(q)$ for a $q \in f^{-1}(B)$. But that $q \in f^{-1}(B)$ means simply that $f(q) \in B$!

Exercise A.10.5   These are just rewritings.

Exercise A.10.8   We have that $p \in f^{-1}(B_1 \cap B_2)$ if and only if $f(p) \in B_1 \cap B_2$ if and only if $f(p)$ is in **both** B_1 and B_2 if and only if p is in **both** $f^{-1}(B_1)$ and $f^{-1}(B_2)$ if and only if $p \in f^{-1}(B_1) \cap f^{-1}(B_2)$. The others are equally fun.

References

[1] Michael F. Atiyah. *K-Theory*. Lecture notes by D. W. Anderson. W. A. Benjamin, Inc., New York–Amsterdam, 1967.

[2] Nicolas Bourbaki. *Éléments de mathématique. Topologie générale. Chapitres 1 à 4*. Hermann, Paris, 1971.

[3] Egbert Brieskorn. Beispiele zur Differentialtopologie von Singularitäten. *Invent. Math.*, 2:1–14, 1966.

[4] Theodor Bröcker and Klaus Jänich. *Introduction to Differential Topology*. Translated from the German by C. B. Thomas and M. J. Thomas. Cambridge University Press, Cambridge, 1982.

[5] Gunnar Carlsson, Tigran Ishkhanov, Vin de Silva, and Afra Zomorodian. On the local behavior of spaces of natural images. *Int. J. Computer Vision*, 76(1):1–12, 2008.

[6] Jean Dieudonné. *A History of Algebraic and Differential Topology. 1900–1960*. Birkhäuser Boston Inc., Boston, MA, 1989.

[7] Michael Hartley Freedman. The topology of four-dimensional manifolds. *J. Differential Geom.*, 17(3):357–453, 1982.

[8] Morris W. Hirsch. *Differential Topology*, volume 33 of Graduate Texts in Mathematics. Corrected reprint of the 1976 original. Springer-Verlag, New York, 1994.

[9] John L. Kelley. *General Topology*. Reprint of the 1955 edition [Van Nostrand, Toronto, Ont.], Graduate Texts in Mathematics, No. 27. Springer-Verlag, New York, 1975.

[10] Michel A. Kervaire and John W. Milnor. Groups of homotopy spheres. I. *Ann. Math. (2)*, 77:504–537, 1963.

[11] Antoni A. Kosinski. *Differential Manifolds*, volume 138 of Pure and Applied Mathematics. Academic Press, Inc., Boston, MA, 1993.

[12] Serge Lang. *Introduction to Differentiable Manifolds*, second edition. Springer-Verlag, New York, Inc., 2002.

[13] John M. Lee. *Introduction to Smooth Manifolds*, second edition, volume 218 of Graduate Texts in Mathematics. Springer, New York, 2013.

[14] John Milnor. *Morse Theory*. Based on lecture notes by M. Spivak and R. Wells. Annals of Mathematics Studies, No. 51. Princeton University Press, Princeton, NJ, 1963.

[15] John W. Milnor. *Topology from the Differentiable Viewpoint*. Based on notes by David W. Weaver, revised reprint of the 1965 original. Princeton Landmarks in Mathematics. Princeton University Press, Princeton, NJ, 1997.

[16] John W. Milnor and James D. Stasheff. *Characteristic Classes*. Annals of Mathematics Studies, No. 76. Princeton University Press, Princeton, NJ, 1974.

[17] James R. Munkres. *Topology: A First Course*. Prentice-Hall Inc., Englewood Cliffs, NJ, 1975.

[18] Stephen Smale. Generalized Poincaré's conjecture in dimensions greater than four. *Ann. Math. (2)*, 74:391–406, 1961.

[19] Michael Spivak. *Calculus on Manifolds. A Modern Approach to Classical Theorems of Advanced Calculus.* W. A. Benjamin, Inc., New York–Amsterdam, 1965.

[20] Michael Spivak. *A Comprehensive Introduction to Differential Geometry. Vol. I*, second edition. Publish or Perish Inc., Wilmington, DE, 1979.

[21] Clifford Henry Taubes. Gauge theory on asymptotically periodic 4-manifolds. *J. Differential Geom.*, 25(3):363–430, 1987.

[22] Guozhen Wang and Zhouli Xu. The triviality of the 61-stem in the stable homotopy groups of spheres. *Ann. Math.*, 186(2):501–580, 2017.

Index

Printed in the United States
by Baker & Taylor Publisher Services